# FACE UP TO CL

DEMAND CHANGE NOW

## OTHER TITLES BY PETER McMANNERS

for Susta Press include:

*Adapt and Thrive: The Sustainable Revolution*
ISBN: 978-0-9557-3690-2

'...governments and businesses who are currently unable to see beyond
the current political term/share prices will oppose this book, but in twenty
years they will be quoting it as the "Essential Guide for Survival".'
ETELÄ-SUOMEN SANOMAT, FINLAND

*Victim of Success: Civilization at Risk*
ISBN: 978-0-9557-3691-9

'This book should be compulsory reading for all people with influence
over the future direction of society, from world leaders and policymakers
to teachers and voters. The time has come to support real change.'
SIR DAVID KING

## ABOUT THE AUTHOR

Dr Peter McManners is an author, consultant, and Visiting Fellow
of Henley Business School. His expertise in sustainability comes from
a multidisciplinary stance including business, geography, engineering
and economics. He works with stakeholders ranging from green
campaigners and activists to business and government. This broad reach
enables him to develop a unique perspective on issues at the nexus of
environmental, economic and social policy. This book brings together
his deep expertise in sustainability and real-world approach with
a radical forward look at a new approach to climate policy.

# FACE UP TO CLIMATE CHANGE

## DEMAND CHANGE NOW

## PETER McMANNERS

Illustrations by THOMAS McMANNERS

SUSTA PRESS
12 Horseshoe Road, Pangbourne, Reading
Berkshire RG8 7JQ, UK

Susta Limited is UK registered company number: 5262675
www.susta.co.uk

First published in Great Britain 2021
Copyright © Peter McManners, 2021
Illustrations © Thomas McManners, 2021

A catalogue record for this book is available from the British Library

ISBN: PB: 978-0-9557-3694-0; EBOOK: 978-0-9557-3695-7

Copyedited by Jill Laidlaw
Proofread by Sophia Blackwell
Project managed by Alysoun Owen Consulting Ltd.
Typeset by Catherine Lutman Design

# CONTENTS

# List of Tables

# List of Figures

# List of cartoons

# List of abbreviations

| | |
|---|---|
| AOSIS | Alliance of Small Island States |
| APD | Air Passenger Duty |
| COP | Conference of Parties |
| CORSIA | Carbon Offsetting and Reduction Scheme for International Aviation |
| IATA | International Air Transport Association |
| ICAO | International Civil Aviation Organization |
| IPCC | Intergovernmental Panel on Climate Change |
| NDCs | Nationally Determined Contributions |
| OPEC | Organization of the Petroleum Exporting Countries |
| PETM | Paleocene-Eocene Thermal Maximum |
| UNCED | United Nations Conference on Environment and Development |
| UNEP | United Nations Environment Programme |
| UNFCCC | United Nations Framework Convention on Climate Change |
| WMO | World Meteorological Organization |

# PREFACE

There should be no need for this book. We should have faced up to climate change decades ago. But we didn't. And now the world faces a crisis.

We are experiencing severe storms, forest fires, droughts and flooding because we burn fossil fuel. Climate change is happening. We know why. We know it will get worse. We know that the consequences could be serious. It would be foolhardy to carry on regardless. The increasing number of national governments declaring a 'climate emergency' are not being over dramatic – this is an emergency which requires urgent action.

The science is well-understood. Although the consequences for a particular place cannot be predicted precisely, it is certain that we are navigating towards dangerous consequences and severe disruption to our way of life. This central message is getting through, and there is a glimmer of hope that the corner might have been turned in dealing with climate change, but nowhere near fast enough. There is the possibility that the climate emergency leads to societal breakdown before we get our act together, collapsing our ability to mount a coherent response. Climate conferences meet but nothing substantial is agreed. The dilemma is solvable if we act quickly and decisively, but not facing up to climate change is perhaps the most extreme example of reckless behaviour in all of human history.

We are now approaching the final stage of the era of fossil fuel. In history this will be recorded as a period when progress stalled. The start of fossil fuel dependency in the nineteenth century will be understood within the context of early industrialization made possible by fossil fuel. Fossil fuel continued to be important through the twentieth century during a period of extraordinary progress. In the twenty-first century, we should now be taking the leap forward beyond the primitive technology of fossil fuel. Instead of building on the gains of the past abandoning outdated methods and developing new capabilities, we seem to have lost our mojo.

Finalising the manuscript for this book in the midst of the COVID-19 pandemic has, paradoxically, given me optimism. During lockdown, people have experienced cities with fewer cars. Those living under flight paths have been able to appreciate less noise. We have all been breathing cleaner air. These are not improvements people want to give back. Facing up to climate change can be about improving lives through reducing reliance on fossil fuel. To start to see the benefits directly is starting to shift attitudes from denial to acceptance, from opposition to support, even per-haps from indifference to passionate advocacy for change.

The evidence is clear, that continuing with fossil fuel is dangerous. We have the means to live without it, but we stubbornly defend an obsolete economic model. Quite why is hard to fathom. Perhaps not enough people open their eyes to reality. Certainly no one alive today knows any economy other than one based on fossil fuel, so we are unable to visualize just how much better off we would be without fossil fuel. Maybe we are simply lazy and self-indulgent. Whatever the reason, this is foolish; we need to wake up and face up to climate change.

I write this book for everyone who cares about the future. This problem is solvable, and the action required is doable. We simply have to get on with it.

*Fossil fuel addiction*

# CHAPTER 1

# TIME TO GET REAL

The world is facing a climate crisis entirely of our own making through relying on the outdated energy source of fossil fuel. We have better technology which can improve our lives, but fear of job losses and lobbying from the fossil fuel industry keeps us trapped in the past. It is irrational to be fearful as it is entirely within our grasp to find a solution. Our fear should be of further delay which would lead to passing a point of no return as the planet transitions to a different climate in a process totally beyond our control. Act now, with urgency, and the worst consequences of climate change can be avoided. Weaning the economy off fossil fuel can be done – as I will explain – but we must get on with it. The most glaring anomaly is continued reliance on coal, the fuel which started the Industrial Revolution two centuries ago. There should be no place for coal in the twenty-first century world economy. It is time that we lifted the economy out of the rut of fossil fuel dependency and launched an economy fit for the twenty-first century, beginning with outlawing coal.

## Addiction to fossil fuel

We are living under the clear and present danger of our addiction to fossil fuel. This is changing the climate of our planet, and it will have consequences. For two centuries we have been burning coal, oil and gas in order to operate factories, travel, stay warm, and keep the lights on. It is hard to imagine life without fossil fuel, so we fail to understand that a better future is waiting to unfold. There is more than enough evidence that our addiction is dangerous; but still we carry on. Our dependency is total and our foolishness almost beyond belief. Of course, it is easier to allow short-term craving to control our actions than to face up to the challenge of curing the addiction. We are like drug addicts lounging around spending their money on the next fix whilst their health deteriorates. We convince ourselves that we feel good, it doesn't matter, and fearing 'cold turkey', we carry on regardless.

Sitting around overdosing on fossil fuel whilst cooking the planet is the behaviour of addiction.

Addiction goes through stages: first, denial of the problem; second, avoidance of the consequences; and finally, acceptance that the addiction is too difficult to cure. The fast track from first fix to the mortuary is a smooth and easy ride. It takes effort and resolve to reflect, slow down, and work out how to switch to a better track. There is no doubt that there is the potential for a great future ahead but our thinking is clouded by our addiction.

## DENIAL

Denial of climate change is the easy unthinking response. Believing it can't be true relieves any anxiety and seems to make the problem go away. For such belief to stick, there needs to be a backstory to explain how climate change became such a big issue. This could be that climate change is a conspiracy dreamt up by scientists wanting to keep their research funding. Such untrue claims can still be found in obscure chat rooms in corners of the Internet; but as the science becomes ever more certain, and the consequences ever clearer, denial is no longer an option. With denial no longer believable, refuge can be found in downplaying significance. It won't be as bad as predicted. The predictions being made by scientists have an element of uncertainty, so whilst there is still a chance that climate change may not be as severe as supposed, there remains a possibility that we do not need to do anything much about it. Some people demand absolute certainty that climate change will have serious consequences before being willing to take action which may change their current way of life. The 'wisdom' of this response is that whilst there is still a slim chance that it won't be a complete disaster, we should wait, rather than invest resources which could be used to satisfy current needs and fix current problems.

Denial of climate change is now rare. Most people accept that the science is accurate; and that the consequences will be significant. The next phase of response is to shift to avoidance.

## STUCK IN LIMBO

Avoidance requires people to be indifferent to the needs of future generations. Some people find comfort in the view that it is not their problem. *It will affect future generations. So, it will be up to them to fix it.* They can also draw on the trend that through modern history each generation has been richer than their parents. *My kids will be richer than me so they can afford to fix what will be 'their' problem.* This is a convenient claim, ignoring the fact that continued failure to face up to climate change will lead to it spiralling beyond human control.

The terminal stage of addiction is acceptance that the cure is too difficult. When it is obvious that the climate is changing, with more extreme storms, hotter heatwaves and longer droughts, people will realize that avoidance has become impossible. The easy route is to decide that it is now too late. Yes, fossil fuel is the prime cause of climate change. Yes, the consequences could be serious. Yes, it is happening now. But it is just too difficult to do anything significant in response. This is where the world is now; stuck in limbo over fossil fuel and climate change.

There is talk about the need for action and talk about what possible action could be taken. We have yet to get real and actually do something. We are now entering an important stage as our deliberations assume a sharper focus taking place whilst we start to experience the consequences of climate change. This will expose the world's current weak approach and open the possibility of changing direction.

# Changing direction

At some point, the various groups within the climate debate should come together and unite; but I doubt this will happen. Climate policy idealists (who talk about fair and equitable responses which will cost little) may unintentionally team up with the climate deniers (who don't care a damn) in a coalition of complete inaction. Unrealistic idealistic aspirations are just as much a block to progress as downright refusal to accept the need to act. To break the stalemate requires a huge dose of realism. We face a pending crisis which will affect us all. A realistic evaluation should lead to tough choices and real action. People who adopt a green perspective need to get real, to see the world as it is, as a competition between nations seeking national advantage to secure their own future. The deniers need to get real, to stop believing that they can defend what they have by simply refusing to give ground. Action is needed because it is vital to our future. Whether you are a selfless person who wants to protect the future of humanity, a selfish person interested in the preservation of lifestyle, or something in between; everyone is just as much at risk. It is time to get real.

Climate change is a crisis, but it is not all doom and gloom. Looking beyond the current crisis there is a bright future. We can be confident that there are viable responses to the climate emergency, not only to deal with the problem but to launch a vibrant new economy with a wealth of opportunities. A major blockage to progress is that this better future is not accessible on the path we are on. To identify real responses, and make them accessible as policy choices, we need a change of direction to open the way to the possibility of a real action. First, we should seek to understand the challenge – fully and in detail – and accept this insight as the foundation for how we think and how we respond. We should not tolerate peddlers of falsehoods, fake theories and lies. We have been slow to get a grip on this. Until recently, it was common practice for news outlets – even credible and respected organizations such as the BBC – to include sceptical views in any piece on climate change, and invite climate change deniers to join with any live debate. It was only when such deniers were limited to a few oddballs from obscure institutions or funded by the fossil fuel industry, that mainstream media realized how the news flow had been subverted to allow doubt to persist in the public consciousness. For decades we have allowed the simple message, that burning fossil fuel is changing the climate, to be confused so that even well-educated professional people doubted this fact, and were persuaded to question the need for action which might have negative short-term economic consequences. With such confusion, it is perhaps not surprising that changing direction is so difficult.

## The core facts

People should be free to express their views and opinions, with healthy debate a key part of the democratic process. But, when facing a crisis, the facts should be rocks embedded in the sands of discussion as fixed points of reference, with ideas allowed to ebb and flow around them. There are multiple possible solutions to consider, ranging from sensible and logical ways forward to off-beat ideas which just might work. Keeping to the core facts allows the best ideas space to grow. The clarity which emerges is incredibly useful. There are two key facts: first, burning fossil fuel is causing climate change; second, to halt climate change we must stop burning fossil fuel. Keeping this simple core logic in mind allows something credible to emerge from complex thought processes and a messy debate.

At a more detailed level, in the flow of debate, some apparently less important facts stubbornly remain like grains of sand caught in an oyster's shell. One of these is that not all fossil fuels are the same. There are relatively cleaner (and relatively dirtier) fossil fuels. All fossil fuels should not be treated in the same way. Each type of fossil fuel should be treated according to its characteristics. This policy pearl is simple and, when explained, completely obvious.

## Clarity and simplicity

The best policies are clear and simple to communicate; the upside is obvious to all, and the downside (there is always a downside) is also clear so that opponents can be clearly identified as vested interests to be faced down. This book presents a number of pearls of policy which can be strung together to make a coherent and credible response to the climate emergency. These are feasible, practical and simple, focused on closing down fossil fuel extraction, with transition arrangements making best use of the cleaner fossil fuels, gas and oil. To prove we are serious, I have then selected the thorny issue of aviation to provide clarity that solutions can be found in every sector – provided we are determined to succeed.

All of us live within a fossil fuel economy, and have done so for many generations. To turn our backs on such a familiar economy is a huge shift. We had a foretaste of the consequences of moving away from fossil fuel when the COVID-19 pandemic gripped the world in 2020. As the world shut down, burning much less fossil fuel, people could sneak a peek behind the curtains of fossil fuel dependency to see a cleaner future.[1] This may encourage more people to share my positive view that taking action which challenges the fossil fuel economy is progress and is not to be feared.

My argument begins with examining the science of climate change. This is the focus of the next chapter – it is short and punchy because the hard work has been completed by the climate scientists and their findings have been interrogated with exceptional rigour. Such rigour is not surprising because the facts are inconvenient – the consequences of accepting them are so influential to policy and are so extensive in terms of the impact on society that we would rather the scientists were mistaken. The climate scientists are not mistaken. Ironically, the huge opposition to the science has served a useful purpose. The science is now rock solid. Where there are uncertainties these have been clearly laid out. There are no uncertainties about the core facts; only uncertainties about predicting the precise consequences. This carefully calibrated uncertainty is integral to the science.

## Dealing with uncertainty

The uncertainty about climate change should not be allowed to disrupt our thinking or delay key decisions. An analogy perhaps helps to explain how it is normal for uncertainty to sit alongside known facts. Let us suppose that a warning is received that a

---

1   Le Quéré et al. (2020), Temporary reduction in daily global $CO_2$ emissions during the COVID-19 forced confinement, *Nature Climate Change*, online 19 May, [available from: https://doi.org/10.1038/s41558-020-0797-x: accessed 19 May 2020].

terrorist group has planted a bomb in a shopping mall, timed to explode in one hour. The terrorist organization has a reputation for accurate warnings and employs a bomb-maker who is known to be an expert. The facts are clear; if nothing is done there will be an explosion one hour from now. The consequences are clear (destruction) but the precise degree of damage is uncertain. No amount of modelling or speculation about the precise nature of the resulting damage can eliminate this uncertainty. The precise damage and cost of the repairs is unknowable. The focus of our attention should be the fact that a bomb is going to explode; there is no need to be distracted by examination of the detailed consequences – especially if this is instead of taking immediate action. Action needs to be based on the known facts. Therefore, the shopping centre should be evacuated and a robot sent in to try to locate the bomb and defuse it. The course of action is clear, even though the authorities do not know the precise consequences, and each shopkeeper doesn't know the precise risk to their business. The bomb is a clear and present danger to the shopping mall and all the shops within it. All shop owners should support the clear and simple plan of action mapped out by the emergency services. Some shopkeepers might want to go in and use the hour to evacuate their most valuable stock, or board up the windows of their shop; but those in control would be correct to set up a cordon around the mall and focus on the plan of action. Selfish individualistic action may even interrupt efforts to defuse the bomb and should not be tolerated. Only after the hour is over, and the bomb has been defused (or not), do discussions about consequent actions have any real meaning.

The uncertainties about the precise consequences of climate change cannot be eliminated. Spending time and expending resources going into more and more detail about possible consequences is simply an excuse for delay and procrastination. The core science is solid. It should be accepted. There is a short window of opportunity to deal with the causes of climate change before serious harm becomes inevitable. This time should be spent focused on action against the causes rather than generating ever more hot air in a debate which swirls around giving an impression that some sort of response is in the wind, when plainly it is not.

## Shifting attitudes

As the science of climate change becomes accepted, there is a discernible and useful shift in attitudes. A growing number of people now accept the need to do something. In a recent poll, a majority of people think that we are still able to avoid the worst effects of climate change but only so long as we take drastic action. This implied support for drastic action ranged from a low of 50% in the United States to over 80% of people in Spain, with the UK in the middle at 66% of those polled.[2] So, the ranks of climate activists demanding action have been joined by a growing silent majority who accept that action will be needed.

This is not yet an unequivocal demand for action, and it is still tempered by worries that action may undermine the economy and affect lifestyle. Action is wanted, but as climate change is not yet accepted in the public consciousness as a crisis, we stick to the condition that any action should not expect too much of us nor put at risk what we already have. This is not yet fertile ground on which politicians can plant a manifesto to deal effectively with climate change. There is too much worry about the

---

2  Smith, M. (2019), 'International poll: most expect to feel impact of climate change, many think it will make us extinct,' YouGov.co.uk, 15 September. Available online: https://yougov.co.uk/topics/science/articles-reports/2019/09/15/international-poll-most-expect-feel-impact-climate [accessed 5 April 2020].

negative consequences for society of effective climate policy, and not enough worry about the negative consequences of climate change. The play-off between this pair of negatives leads to paralysis in public opinion. Introducing a positive outlook could provide the tipping point which quickly, and perhaps abruptly, alters the political landscape. The positive perspective that I believe can change the balance is the realisation that effective climate policy has overall positive outcomes for society. The changes required can be welcomed for the improvements they bring, even if climate change is not as dangerous as predicted. Of course, we know now for sure that climate change is a very real threat; and the change of language is helpful when referred to as a 'climate emergency'. When a promise of positive outcomes is linked with a clear understanding of the possible consequences of not taking action, this produces a combination capable of attracting widespread support.

An unexpected benefit of the recent COVID-19 health crisis has been the way it has provided a nudge towards the acceptance of climate action. Restricting the spread of the virus closed down much of the world's fossil fuel transportation. People could see and experience the world becoming a quieter and cleaner place. Aviation was closed down, and people living under flight paths experienced the improvement this brought, while business people learnt to make use of virtual conferencing as an alternative to travel. The airline industry has been decimated in much the same way as one would expect of effective climate action. Rather than try to get the aviation industry back to where it was before COVID-19, this opportunity should be used to relaunch aviation in a different climate-friendly direction (more on this in Chapters 10 and 11). Another perhaps more persuasive perspective arose from the impact of the transport lockdown on cities. The city of Mumbai is a prime example of a place where people had been breathing some of the most polluted air on the planet. During the pandemic lockdown people could see and feel the health benefits of clean air. They may not want to go back to how it was.

The consequences of COVID-19 hit the world's economies hard. The heavy lifting required of government for recovery is an opportunity to construct a climate-friendly economy. Instead of investing to get back to where we were, we should invest to set off in a new direction. As the world picks itself up beyond the pandemic, there is a golden opportunity to align our actions with facing up to climate change.

The facts of climate change have been known for decades, but it is only recently that the circumstances have been shaping up for effective action. It is human nature to delay and wait to be forced to act. All sorts of risks and dangers demand resources, but resources are not allocated until the risk or danger is clear and present. Now we know that climate change is a clear and present danger to human civilization as we know it. Whether you agree or disagree with this statement, I urge you to read the next chapter on climate change carefully. If you agree already with my assessment of the importance of climate change, it will focus your attention on the key parameters to allow you to think about what to do and arguments to deploy in conversation with others. If you disagree with my claim, that civilization is at risk, you will find it an interesting chapter to open your mind to possible future consequences which you might not have considered. I do not expect to carry everyone with me. I have a friend with whom I have robust arguments. He argues that it is not for him to accept the cost and take responsibility for something which will impact mainly on the generations to come after he has died. He has already suffered one heart attack, so perhaps already takes a short-term view of his own life and wants to live out his remaining years without disruption to his lifestyle. For other people, the realization that a crisis is pending, and that climate change is dangerous, breaks their resistance to inconvenient facts, allowing them to start to embrace the possibility of taking real action. My friend, and those of a similar mindset, offer the world a triple whammy. First, don't accept any responsibility; second, defend what you have; and third, insist that others pay the price. Sadly, in the modern world, this seems logical and quite normal. This allows you to get sucked into a fantasy that climate change really doesn't matter and, in any case, nothing much can be done about it. This is not a fantasy. This is real. This is solvable. If we get real, it can be solved. This will require investing time and resources. The justification is simple. We are facing an existential crisis to which we must respond.

Getting real about climate change has a number of elements, including understanding the nature of the challenge; accepting that it could be serious; shifting attitudes from indifference to engagement; and calling for leaders in all walks of life to initiate action. Now that the risks are clear, and the response so far has been exposed as inadequate, the time is ripe to look for and bring on the tipping point in public attitudes. Above all else, getting real at this particular stage in world affairs requires that people get angry, very angry. Not just a few activists or environmentalists but an upswelling of anger from everyone, in all places, of all ages. Politicians can react very fast in the face of widespread public anger.

The potential for civil unrest, and the threat of a breakdown of law and order, are powerful levers for change; but only when the action required of the politicians is clear and precise. In the area of policy, examples could be: a policy which needs to be adjusted, an existing policy which should be withdrawn, or a new policy which should be implemented. The clarity of the demand which fuels public anger is as important as the breadth and depth of the public rage. Clear demands married to evident anger get the political class very worried. When politicians are handed the changes that need to be made on a plate, and the changes make sense to the majority of people, politicians will quickly embrace and implement them to defuse the situation before it escalates.

This book provides clarity of the demand for a response to the climate emergency. I encourage everyone who reads it to contribute to a groundswell of anger expressed in whatever way they feel could have impact and get noticed, as individuals and as groups. The variety and intensity of how the anger unfolds is all part of showing politicians that, whatever ambivalence they might feel towards climate change, there is a growing social movement which will unseat them and drive them from power unless they respond.

Before using your power and influence (which as part of a greater movement can be very great indeed) you need to get familiar with what to demand. This starts with really under-standing the facts about climate change. This is the logical place to begin getting real. When you can see that this is not an impos-sible problem, but one which can be solved, it is inspiring. All that is needed is to find a common sense of purpose and the cour-age to get on with it.

*The Reckless Experiment*

# CHAPTER 2

# CLIMATE CHANGE
# – THE ESSENTIAL FACTS

This book is about responding to climate change by eliminating fossil fuel dependency. This will mean closing down fossil fuel extraction (starting with the dirtiest coal) and building a cleaner economy. These are positive steps forward, to be welcomed for the improvements they bring. The climate emergency provides the impetus to get on with it. If you already know about the dangers of climate change then you can skip this chapter. However, climate change is the reason why eliminating fossil fuel dependency is now urgent, so this chapter is a necessary step in the logic of arguing for bold and rapid action.

Discussion of climate change – and the debate about what to do about it – has become overcomplicated by trying to make precise predictions. The insistence on working out exactly what will happen is a trap, providing the excuse for procrastination. Rather than get bogged down by complexity, it is useful to cut through to focus on the essential facts.

Climate scientists have put considerable effort into understanding the Earth's complex climate system. For the rest of us, a simplified model is enough to understand the essential parameters, starting with the prime source of energy which drives the system, the Sun. The Sun shines on the planet; some is reflected (particularly off clouds and areas of white ice); some heats the atmosphere; some heats the land; some heats the oceans; and some is radiated back out through the atmosphere into space. Atmospheric winds and ocean currents move this heat around from one place to another. The swirling winds take up moisture over the sea and when the air cools (such as arriving over land and rising to higher altitude) the moisture condenses and falls as rain. The rain runs off the land, finding its way into rivers and back into the ocean. Some of this water falls as snow and its journey back to the ocean is delayed, captured as ice at the poles and in the high mountains. Seasonal snow comes and goes, but over many hundreds of thousands of years snow has accumulated in vast quantities, compressed into ice sheets covering areas around the poles and as glaciers in the mountains. In a nutshell, this is the climate system. This is all that is needed to understand the essentials of the climate system, apart from one other important component, the role of carbon dioxide.

# The role of carbon dioxide

If levels of carbon dioxide remained constant, we could happily ignore it as a factor. Up until the Industrial Revolution levels of carbon dioxide in the atmosphere had been stable for hundreds of thousands of years. The weather systems of the planet, although hugely complex, had settled into a predictable and stable climate. In the short term, from year to year, there would be fluctuations and extreme weather events. Looking back through the historic record, it was possible to describe extreme weather as once-in-50-years, or once-in-100-years events. There was no need to understand the complexities of climate science. From weather records we knew what weather could apply in a particular region. The likelihood of any particular weather event (and its severity) could be found in the statistics. That is how it was before carbon dioxide levels rose significantly above what had been normal.

Since 1800, levels of carbon dioxide in the atmosphere have been climbing; at first slowly and more recently ramping up steeply. The level of carbon dioxide in the atmosphere was 280 ppm in 1800 and has now exceeded 400 ppm. Consistent and accurate direct measurements have been available since 1958 when the Mauna Loa Observatory, Hawaii, started continuous monitoring. The steep rise in $CO_2$ is one of the known facts which anchors our analysis. It is no coincidence that the climb in $CO_2$ concentrations coincides with the Industrial Revolution, powered first by coal and then later by other fossil fuels. There can be no doubt that this rise in carbon dioxide concentration is a direct consequence of burning fossil fuel. It is therefore very important to understand how our reliance on fossil fuel is changing the climate.

# Global warming

The key to understanding the role of carbon dioxide is that it is a 'greenhouse gas'. This means that it limits the amount of heat radiating back out to space through the atmosphere. Higher $CO_2$ levels mean that the Earth loses less heat, so in the Earth's closed system in the vacuum of space, it warms up. The higher the concentration of carbon dioxide in the atmosphere, the higher will be the average temperature of the Earth. This is global warming. Again, this is factually accurate and backed up by data. The 2010s was the hottest decade and at the time of writing, the last six years were the warmest on record.[3] It may seem unnecessary to be so careful in building the logic but it is important to know what are facts (which we know for certain) and possible consequences (which are open to interpretation). It is a fact that burning fossil fuel is warming the planet. Most people reading this paragraph will be thinking this is obvious, so why state it? The reason is that there have been people,

---

3   The warmest years were, in order: 2016, 2019, 2015, 2017, 2018, 2014. Source available online: https://www.climatecentral.org [accessed 26 July 2020].

and there are still a few, who doubt this fact – or rather seek to sow doubt about it, because they don't like the implications which follow on from it. Our actions in burning fossil fuel are warming the planet and if we want to avoid the consequences of a warmer planet, we have to stop burning fossil fuel. Simple irrefutable logic, but leading to complex consequences for both society and the economy.

The possible consequences of warming the planet are subject to much detailed analysis and debate. The climate system has appeared to be consistent through most of human history, acting as a stable backdrop for society. In the twenty-first century, this is no longer the case. Two centuries of burning fossil fuel has changed the composition of the atmosphere; as a consequence, the climate has started making the transition to a different climate. We might wonder why has it taken so long to recognize the danger. The reason is that there is considerable delay in the system. Until recently, the climate had not changed enough to be obvious. It takes time before climate change shows up in the aggregated weather measurements. There will also be delay in slowing or halting climate change after we stop burning fossil fuel. When we stop releasing fossil carbon, the transition will still continue for many decades, or even centuries, into the future. If we stop messing with the climate system soon – by the rapid elimination of fossil fuel from the economy – we will still have to live with the changes already pent up in the system. If, instead of taking action, we continue to release fossil carbon into the atmosphere then the consequences will be increasingly severe, and the outcome beyond the twenty-first century devolves into scary scenarios which are hard to contemplate. It seems impossible because we have climate records going back over history and nothing really terrible has happened in the recorded past. We conveniently forget that all historic records have no relevance now as the Earth adapts to levels of carbon dioxide not seen for a million years.

# Forecasting the consequences

Forecasting the consequences of climate change needs great care, especially without historic records to help. In all of human history we have not been here before. What we are doing is carrying out a reckless experiment on planet Earth.

When scientists carry out theoretical experiments by building climate models inside computers, the numbers are crunched based on a set of assumptions. These generate possible climate scenarios. These are not reality, of course – no matter how scary the scenario might be, it can be erased by hitting the delete key. Perhaps this is why disturbing theoretical scenarios presented by climate scientists tend to be ignored. It seems that predictions of dire consequences tend to close people's minds, make us inclined to ignore the message and prompt the question, 'Where is the proof?'. People are withholding support for policies which might affect lifestyle or economic situation, until it can be proven that the world is on its way to climate meltdown. Of course, the scientists have no proof; all they can provide is their expert judgement and theoretical scenarios. The key point to understand is that there will be consequences. We live on planet Earth and totally rely on its ecosystem. We will experience the results of the Reckless Experiment without any way to control it. There is no delete key we can hit to escape reality. When you think deeply and clearly, we do not need precise predictions. This Reckless Experiment simply should not be running.

Engaging in the prediction game is dangerous. Our participation implies that we are willing to accept a certain elevated level of carbon dioxide in the atmosphere up to which the Reckless Experiment can continue – dependent on the severity of the predicted consequences. As predictions are made, they can be countered by 'Where is the proof?'. But there is no proof;

there can be no proof; there will be no absolute proof. So, by demanding proof, we migrate away from the clear logic that this is an experiment we should not be running – to leaving it to run while we consider possible consequences. The warped logic which can be employed is that we don't have solid proof about the consequences so we do not have grounds to stop the Reckless Experiment. Solid proof will only come as we complete the experiment. We want to know the consequences with a high degree of certainty before we will invest time and resources to stop the experiment; but until the experiment is completed, we cannot be sure of the consequences. Add to this the fact that the final end of the experiment is centuries away and we are trapped. The way to escape this trap is to return to the beginning and adopt the clear logic that the experiment should not be running.

So, I have written all that needs to be explained about climate change. The trouble is that stopping the Reckless Experiment will not be easy.

## Stopping the Reckless Experiment

Ceasing to burn fossil fuel has its own set of consequences. These may not have the scale, importance or impact of the Reckless Experiment, but they are more immediate and more personal. Stopping burning fossil fuel means closing down the fossil fuel economy and launching a different economy. The potential impacts of this are direct and personal, so in many people's minds seem more important than the risks of climate change. These short-term concerns mean that it is necessary to dwell a little more on climate change, to tease out further insights, so that even the most selfish person with the shortest time horizon can understand the need to take action.

We may have no way to know the precise consequences of climate change, and have no historic records which might help, but we do have the geological record of the planet. From this we know that the planet in the distant past has had high levels of carbon dioxide in the atmosphere. One such period in geological time is termed the Carboniferous period. This is when the planet was hotter and wrapped in dense forest. Over millions of years, the trees grew, died and decayed, laying down the deposits which were then transformed through heat and pressure into coal, oil and gas. This process took carbon dioxide out of the atmosphere and locked it away underground – this is the same process we are now reversing by burning fossil fuel. As carbon dioxide levels dropped, the planet's climate adjusted to be cooler and settled into a low-carbon dynamic equilibrium. The cooler Earth started to accumulate ice at the poles and in high mountain areas, thus taking water out of the oceans in such huge quantities that global sea levels reduced by more than 60 metres. Again, this is something we are seeking to reverse within the Reckless Experiment. A total reversal which melted the polar icecaps and all the glaciers would return sea levels to over 60 metres above current sea levels. That should be food for thought. It puts my house, located 50 miles (80 kilometres) inland from London, on the edge of the new coast of the UK. Most of what is now Greater London would be submerged, apart from an archipelago of islands including Hampstead, Harrow and Highgate.

## The possibility of abrupt climate change

There is one particular episode recorded in geological history which is particularly interesting and pertinent to our current circumstances. This is the Paleocene-Eocene Thermal Maximum

(PETM) which occurred about 55 million years ago. There was a large injection of $CO_2$ into the atmosphere, increasing global temperatures by 5–8 degrees, leading to a warm period of approximately 200,000 years.[4] Exactly what caused this is an active debate among scientists and academics. One theory is based on methane hydrate. This builds up under the seabed alongside the processes which make fossil fuels. Unlike fossil fuels, which are locked away deep in the Earth's crust, methane hydrates are volatile. When they warm up the potent greenhouse gas methane is released into the atmosphere. The theory goes that as the methane was released this caused further warming, leading to further methane release in a reinforcing cycle of climate change. This theory is relevant to today because there are extensive stores of methane hydrate on the seabed and under the type of vegetation common in polar regions called tundra. All it needs to get released into the atmosphere is for it to be warmed up. Humankind's actions warming the planet could be enough to destabilize methane hydrate to initiate runaway climate change. This is just theory. There is no proof, but it does make us reflect and think.

The geological record shows that abrupt climate change does happen. Two centuries of sustained carbon dioxide emissions by human civilization have caused an abrupt change to the composition of the Earth's atmosphere. It is not too far-fetched to suppose that the continued burning of fossil fuel could cause runaway climate change. Instead of a process taking centuries, as predicted by carbon dioxide emissions alone, it could happen even faster and be more severe than predicted. This insight should be a worry to everyone; but there is no proof. Why do

4    Bowen et al. (2015) 'Two massive, rapid releases of carbon during the onset of the Palaeocene–Eocene thermal maximum' *Nature Geoscience* 8, 44-47. Available online: www.nature.com/articles/ngeo2316 [accessed 26 July 2020].

we keep demanding proof of exact climate consequences when the risks of running the Reckless Experiment are so great? I do not claim that the world's climate is about to rapidly run out of control, but if our actions set off such a mechanism, we will not be able to stop it. We have the choice, now, to stop burning fossil fuel; but we will not have the choice in the future to backtrack if we find the climate changing rapidly. It will change until it reaches a new equilibrium; which will be what it will be. No one can predict accurately where the climate will end up.

## We know enough

Having dipped into possible scary outcomes, let us return to that which we can predict. The planet will continue to warm. Continued warming is an accurate prediction which we can take as fact. There will be disruption to agriculture, storms will be more frequent with greater severity, and sea levels will rise. As a consequence, food supplies will be disrupted and there will be a growing number of climate refugees. These are the circumstances under which conflict and wars are likely, as people who draw a losing ticket in the climate lottery fight for survival. As previous weather patterns are overwritten the precise way these consequences play out for each region and locality is uncertain. The core fact to keep uppermost in our minds is that releasing fossil carbon is changing the weather system and the precise consequences are unknown. For something as vital to our ability to live, grow crops, and remain safe from extreme weather, it is sheer madness to press on with relying on the fossil fuel economy. There is very little need, or any point, in asking climate scientists to sharpen their predictions. First, they can hardly do this; second, we already know all that we need to know. We know that the climate is changing; that increasing levels of carbon dioxide are the prime cause; that we are responsible for causing it. It is therefore our responsibility to stop climate change – except that we can't. The changes to the climate which are already underway cannot be stopped; but all is not lost (yet). We can halt further release of fossil carbon to prevent further man-made pressure on the climate.

# Turning a deaf ear

The knowledge that the climate is at risk is not a new discovery. Scientists have understood the mechanisms and have been predicting likely outcomes for many decades. My own initiation into climate science was in the early 1990s. I was studying for a Master's degree in Geographic Information Systems at the University of Edinburgh. For my dissertation, I chose to examine global environmental databases and the need for ground-truth data to complement the data from earth-observation satellites. As part of my research I examined the uses to which global environmental data were being put. I found that within the scientific community of environmental and climate experts it was well known, and well understood, that human activity was changing Earth's systems. I was astounded that this knowledge seemed not to have been shared more widely. When I investigated, I discovered that the scientists were not keeping the knowledge under wraps; they were very keen to share their findings, but the world was offering a deaf ear. The message that humankind was messing with Earth's systems, and that the consequences could be dangerous, was not a message people wanted to hear. So, the insight that long-term environmental damage was a consequence of our prosperous lifestyles was not a concept the wider community of people and policymakers wanted to embrace. The science of climate change was not ignored entirely at the time. Discussions facilitated by the United Nations were taking place, culminating in the formation of the United Nations Framework Convention on Climate Change (UNFCCC) – more on this in the next chapter. Such discussions have provided (and to an extent continue to provide) warm reassurance that action is in hand, allowing policymakers to carry on and not worry too much about climate change.

After completing my Master's degree, I joined the ranks of the people who knew something was up but were not overly concerned. It was over a decade later before I had my light-bulb moment which woke me out of my indifference.

# Melting the Greenland ice sheet

In 2005, I was living in Helsinki and writing a book about the 'Sustainable Revolution'.[5] The book envisaged a revolution taking shape with sustainability starting to drive policy. Climate change was one of the factors making sustainability important but it was not central to my thinking. I was thinking through the whole range of society's responses to environmental challenges and how these responses will shape a new and better economy. It was clear to me then, and in even sharper focus as I write now, that the sustainability agenda is important to secure the future of humanity. So, sustainability matters; but I wondered if climate change was a peripheral issue or a core challenge which frames sustainability. I wasn't sure; a change in climate will impact some people, some places, and some regions; but the outcome could be both positive for some and negative for others. I admit that my view back in 2005 was that climate change is happening, but of all the challenges we face, it did not stand out as deserving special attention – until a meeting in the British Embassy in Helsinki.

---

5    McManners, P.J. (2008), *Adapt and Thrive: The Sustainable Revolution*, Reading: Susta Press.

The meeting had been set up by the embassy to bring together experts and policymakers from the UK and Finland to discuss climate change and its implications. One of the speakers was from the UK Meteorological Office. The presentation made me sit up. The focus was on Greenland, a massive island in the North Atlantic under Danish control which is covered by a huge ice sheet. This has built up over many hundreds of thousands of years of snowfall and in places is over 4 kilometres thick. The logic explained was this: We know the precise area of the ice sheet and using radar we can accurately measure its thickness. It is simple arithmetic to calculate how much water is trapped on Greenland. It is only marginally more complicated to work out the sea level rise which would result if the ice sheet were to melt entirely. Such calculations show that if the ice sheet melts, global sea level will rise 5–6 metres. This is a hard fact which cannot be disputed. The meteorologist then presented their judgment that the world is close to the point at which the level of $CO_2$ in the atmosphere will lead to sufficient warming to melt the entire Greenland ice sheet.

Sitting in the plush meeting room of the British Embassy, it was easy not to feel at risk. The meteorologist was not saying that all the ice on Greenland would be gone in the near future. People in the audience, who were more interested in the short-term and immediate consequences, could relax and sink lower into their comfortable seats. Meanwhile, I sat bolt upright. It was further explained that the entire melting of all the ice on Greenland would perhaps take up to two centuries. The insight which had such impact for me was that, once the tipping point has been passed, there will be absolutely nothing that can be done to stop it. The pictures we see each summer of torrents of water funnelling down through the Greenland ice make this possibility seem very real. Is it just a possibility that all the ice on Greenland will melt or has the tipping point already been passed? We have no way of knowing for sure.

Over a decade after that meeting of experts, the possibility that the Greenland ice sheet has passed its tipping point is ever more likely. Recent research published in the journal *Nature* in December 2019 stated that although the ice sheet was close to a state of balance in the 1990s, it is now melting fast. Using satellite measurements of changes in the ice sheet's volume the researchers concluded that Greenland lost 3,800 billion tonnes of ice between 1992 and 2018, causing the mean global sea level to rise by over 10 millimetres.[6] Pause and consider the consequences. This indicates that we may be accelerating towards the point of no return. Knowing that sea levels will be 5–6 metres higher two centuries from now has profound implications.

I like to get a feel for such a prediction in my own mind. Unlike Antarctica, which is a continent and likely to hold onto its ice for some time, the region around the North Pole is a sea and the ice is already melting at an alarming rate. Open water at the North Pole during the summer is expected at some point in the next few years. The significance of this is that instead of white ice reflecting much of the Sun's heat back out to space, the dark waters of the sea will soak it up. This means that polar regions will warm rapidly as heat is pulled into the Arctic Ocean. Greenland is now part of the frozen Arctic region, but in the near future it will be surrounded by relatively warm water. It is entirely credible that once the North Pole is ice-free (that is now inevitable) it would follow that the melting of the Greenland ice sheet would accelerate and be unstoppable. There is no proof that this will be the outcome, but looking from an alternative

---

6   Shepherd et al. (2019), 'Mass balance of the Greenland Ice Sheet from 1992 to 2018' *Nature* 579, 233–239. Available online: www.nature.com/articles/s41586-019-1855-2 [accessed 26 July 2020].

perspective, claiming that the Greenland ice sheet could survive in a warmer world seems indefensible.

The future for Greenland could be bright. It is thought to have considerable mineral resources which could be exploited when the ice retreats. Instead of a few small settlements living a simple existence on the edge of a frozen wilderness, Greenland could become an industrial and agricultural powerhouse as its land is exposed for the first time in human history. Of course, this would only be land 5 metres or more above sea level. In Greenland (and everywhere around the world), land below 5 metres elevation will have to be abandoned as places to live, or to farm, as sea levels rise.

# The risk to coastal cities

The sledgehammer insight for me, which brings it very close to home, is the impact that rising sea levels will have on our cities. Sea defences of the scale required to defend coastal areas would simply not be feasible; the amount of resources would be astronomic. Perhaps because of the long timescale, we are fooled into thinking that it's not going to happen in our lifetime, so it's not our issue.

Cities are designed and their infrastructure built and maintained according to long-term strategic plans – or at least they should be. In less developed poorer countries, cities evolve and grow in a haphazard manner. In the developed world, we pride ourselves in doing better than this and take a strategic approach. The insight that sea level rise is now unstoppable, and could be many metres, is profound. The heavy concentration of population and assets in port city locations means that the consequences of sea level rise will impact on many people and have severe economic consequences. The top ten countries most at risk include the Netherlands (as you might expect) with the top three countries at risk being China, the United States and India, in that order.[7] If, as seems likely, Shanghai, Manhattan, Mumbai and many of the world's great coastal cities are to be submerged in the next two centuries, alternative cities will need to be planned. As the City of London is lost to the sea, the UK will have to nominate a new capital designate. These replacement cities will need to be sited on higher ground and long-term infrastructure investment redirected from the old great cities to these new cities of the waterlogged future.

Once governments start to understand the pending future for coastal cities, and get a grip on the issue, they will withdraw investment. The decline of coastal cities will therefore come much sooner than the sea takes over. These are not decisions for some remote future, but strategic decisions required to be taken within a decade or so. The full consequences may be centuries away, but just a few decades into the future, those people in power who have deliberately chosen to ignore climate change may yet get to see the early results of their incompetence.

Through the twentieth century the global sea level rose by about 20 centimetres. The sea level rise predicted this century through to 2100 varies from about 50 centimetres to 1 metre. For someone living in the UK, this does not seem too bad. Other people, in other countries, would disagree. Low-lying island states such as the Maldives will simply no longer exist. We should keep in mind that the world will pass tipping points, when the next ratchet up in sea level becomes unstoppable. I have used Greenland as an example; there are also the glaciers in the

---

7   Nicholls et al. (2007), 'Ranking of the world's cities most exposed to coastal flooding today and in the future.' Available online: https://climate-adapt.eea. europa.eu/metadata/publications/ranking-of-the-worlds-cities-to-coastal-flooding/11240357/view [accessed 5 April 2020].

high mountain ranges including the Himalayas, the Andes and Alps; and the biggest block of ice on the planet, Antarctica. These will not melt in sequence, one after the other, but in unison as every frozen corner of the planet drains down into one global ocean. Melting glaciers will have other less predictable impacts. For example, the Himalayan glaciers act as a water tower which keeps some of the great rivers of Asia in steady flow between the monsoons. Without them, rivers might cycle between flood and muddy trickle. I believe that there is no need to go into the full range of possible consequences of climate change. The single focus on sea level is reason enough to take action. Paradoxically, adding in more potential impacts of climate change, although strengthening the logical argument, can weaken human resolve. When everything looks hopeless, we can become overwhelmed and our minds struggle to cope.

Continuing the focus on global sea levels, there is an upper limit which cannot be exceeded of 60–65 metres. If the world continues with fossil fuel and pushes temperatures to the point where the planet has no permanent ice, it is a plain fact that an ice-free planet would have a sea level 60–65 metres higher. There would be little comfort in celebrating that Greenland and Antarctica become new areas of land to occupy when so much other land had been lost. The sea level rise of 60–65 metres is not a rough estimate – it is known with accuracy and certainty that the shoreline of an ice-free planet would be higher than the 60-metre contour shown on current maps. An interesting exercise for everyone is to take a look at a local map and read off the elevation of their house and community. If it is below 60 metres, it is at risk from being inundated by the rising sea at some unspecified time in the future if we fail to cure fossil fuel addiction.

I have focused on one consequence of climate change – sea level rise – because it is an easy hook to comprehend the scale of the challenge. Whether you find the particular example of Greenland as compelling and hard-hitting as I do, you have to acknowledge the fact that melting the Greenland ice sheet would submerge much of the world's coastal cities. Add to this clear fact the observation that the world has done nothing to back off from burning fossil fuel, and it makes the melting of Greenland ever more likely, and the impact on coastal cities ever more certain.

## Other consequences

There are many more consequences of climate change in addition to low-lying land being inundated by rising sea levels. Agriculture will change as farmers are forced to grow different crops to suit a different climate. Some areas which are currently good agricultural land may become desert. Instead of the usual climate being a stable factor for a region or place, it will become a lottery in which changes may be beneficial or negative. These geographic changes will have been initiated by our actions – burning fossil fuel over a long period of time – but such climate consequences will be entirely out of our control. We will have to respond; to relocate; to adopt a different lifestyle; close down whole communities (or even countries, as with the Maldives) and move to some other part of the planet (if those who live there will accept incoming climate refugees). These human consequences would be hugely challenging, certainly disruptive, and likely to lead to conflict between people and between nations.

## People's attitudes

Having outlined the importance of climate change, and discussed the potential severity of the consequences, the human aspects are interesting – and disturbing. We know our actions are causing climate change and soon enough we will see the direct impacts as the global climate shifts direction. I find the attitudes of some people really intriguing – the ones who have deeply ingrained in their psyche that they don't want to know about climate change and they don't want to have to do anything about it. Quite how they can be happy to be in such a place is hard to pin down. The closest I can get to understanding this resistance is to return to the analogy which runs throughout this book, that we are suffering from fossil fuel addiction. When you examine this human reaction to climate change as an addict's attitude to their addiction, all becomes clear. Rational arguments to halt fossil fuel usage do not do the trick, and are simply ignored. There needs to be an impassioned and hard-hitting explanation of the overwhelming imperative to change behaviour, which hits home to the addict. The addict must face up to the situation, but this takes courage and commitment, which in addicts tends to be in short supply.

The nature of the psychological trap was brought home to me during a public meeting where I was one of three keynote speakers. I spoke about sustainability and the need for policy to ensure a sustainable future, where environmental protection is integral to how society operates. The audience seemed to be receptive to my message and I received warm applause. The second speaker was a climate expert, whose style was a little dry but he was precisely correct in his explanation of how the climate is likely to change, being careful to state what is fact and that which is less certain. As an explanation of climate change and potential impacts it was exemplary. He received polite applause. The third speaker was Christopher Monckton, a climate sceptic and regular public speaker opposing action to deal with climate change. He spoke with force and passion, presenting his 'analysis' that climate change was not a serious threat and wasting resources to respond is unjustified. His argument was that all this talk about climate change was scaremongering. I expect that as people become increasingly savvy, they won't stand for such bunkum, but at this meeting some years back Christopher Monckton received rapturous applause. This was an audience of educated people, who must have known deep down that this couldn't be correct. However, it was so much easier to believe the message that climate change isn't happening, or if it is, it hardly matters. It seems that they wanted to believe Lord Monckton, in order to be able to blank out the warning message from the climate scientist. Afterwards, talking with the climate scientist, he was shocked that such misrepresentation of information could be spun so skilfully to capture an audience. I was bemused to experience firsthand the human capacity to allow a convenient but odd belief to override hard logic. This human trait is what conmen use to ply their trade. We are naturally predisposed to accepting a message which benefits us, and to blank out messages which don't. Whilst there are still people who argue conveniently that climate change is not important, we don't have to respond. As respected climate deniers quietly leave the stage (as surely, they will), leaving only people who are clearly oddballs, or unqualified 'scientists' funded by the fossil fuel corporations, we should wake up to the con.

Another interesting aspect of the human reaction to climate change is the nature of discussion over how to respond. There are people who take the view that it is not their problem. They feel confident that they can live out the rest of their lives before consequences really hit home. Even when it is explained that it will be their children's problem and definitely a problem for their grandchildren, they manage to retain the attitude that it is not their concern. It is not my problem; it is not our problem; it is a problem for future generations. I find such lack of concern deeply troubling. Perhaps it is a consequence of the modern world encouraging the belief that material wealth, here and now, trumps the concept that we are but a small part of the human project which has roots in history and a future yet to unfold.

I look to everyone to re-examine our 'modern' attitudes, by looking first into ourselves and then talking with and influencing others. If you are a young person, then climate change will hit you in your lifetime. If you reflect on what your elders have done, and how little they are doing now to correct their mistakes, it will not be surprising if you are angry, very angry. As I was completing the manuscript of this book, Greta Thunberg, the sixteen-year old Swedish climate activist, was named *Time* magazine's Person of the Year 2019. Fortunately, people like her are waking up in increasing numbers and realizing that climate change is a clear and present danger. I am ashamed that it took so long, but we are where we are. If there is any consolation, it is that these decades of denial mean that now we know firsthand that climate change is real. The evidence is there to see, as more extreme weather events and climate records are broken across the planet. Let us not focus on the past and focus instead on waking up to reality and leaping into action.

## Conclusion

Climate is a complex interaction between a number of factors including incident energy from the Sun, winds in the atmosphere, ocean currents, clouds and all manner of other influences. We don't need to understand the full complexity to understand climate change. In the past, we could simply measure and record the weather to build databases of climate statistics. Weather patterns tended to repeat year after year, and extreme weather events occurred with the same probability as shown in the historic records. This worked because the world's climate had settled into a stable equilibrium. The assumption that climate is stable has served humankind well enough over the centuries, with good years and bad years, but overall we got the climate we expected based on weather records going back over many decades. The problem we face now is that the parameters of the climate system are changing as the planet warms.

Scientists seek to predict the consequences of climate change but in truth the task is beyond them. Our Reckless Experiment is pushing carbon dioxide in the atmosphere to levels far in excess of pre-industrial records. This will certainly shake up the climate system, but what sort of new equilibrium the system adopts is hard to predict. It is not simply climate which is at risk; biodiversity loss will accelerate as habitats change, so we might face total ecosystem collapse. It will take many decades or perhaps even centuries before a new stability is established. We may have to live with a chaotic climate system, and the associated severely degraded ecosystem, for many generations into the future unless we act decisively and quickly. All we know for sure is that increasing levels of carbon dioxide are changing the planet's climate; but that is all we need to know. To deliberately change the climate, without knowing the consequences, would at the very least be unwise. Finding out that your actions are changing the climate, although this was not your intention, and then failing to call a halt is equally indefensible.

There is only one way to slow and then halt global warming – slow and then halt the emission of fossil carbon into the atmosphere. And there is only one way to slow and then halt the emission of fossil carbon – stop burning fossil fuel. The complexity and uncertainty of climate change leads to a simple and clear conclusion that we must close down extraction of fossil fuel commencing with coal.

*Life beyond fossil fuel*

# CHAPTER 3

# DECISION TIME

It is now decision time to face up to climate change. It is important to find the wisdom to make the right choices, the courage to defend them, and the determination to follow through with action.

The key decision is about carbon dioxide levels in the atmosphere. This is the cause of the problem so this is where we need to focus. There are two broad categories of action: first, stop burning fossil fuel so that carbon dioxide levels stop rising; and second, strip carbon dioxide back out of the atmosphere. The first is straightforward. The second is more complex. There are technologies which could be deployed to remove $CO_2$, but the scale required to make a difference makes this a doubtful proposition. We can keep in mind for the future that affordable ways to sequester carbon dioxide could be a method to deploy, but let us not expect too much. Especially, let us not delay our first response – of closing down fossil fuel consumption – in the hope that in the future we may be able to strip the carbon dioxide back out of the atmosphere.[8] It is simple common sense that it is bound to be easier to stop releasing fossil carbon than it will be to recapture it. Stopping burning fossil fuels is relatively easy; stripping $CO_2$ back out requires technology and investment. It should be a 'no-brainer' to find the collective resolve to stop burning fossil fuel.

---

8   Extensive planting of forests should be part of our long-term response to slowly draw carbon dioxide out of the atmosphere in an affordable way but there should be no pretence that this is the solution. This is part of the long-term recovery process after implementing the solution. The solution has to be to wean the world off fossil fuel.

There seems to be deep reluctance to wean ourselves off fossil fuel. Why people resist is hard to pin down. Part of the reason is detachment from the dirty end of the process. For example, power stations are generally in someone else's backyard and car exhaust pipes are low down at the back, out of sight and out of mind. Another aspect is a tendency to associate the elimination of fossil fuel with a frugal lifestyle. Because the modern world is so reliant on fossil fuel, it is assumed that eliminating it would undermine our lifestyle. How wrong that is. We have the technology to do everything we do now without fossil fuel. Generating electricity, keeping buildings warm, even the apparently impossible challenge of flying without conventional fuels derived from oil, can all be accomplished. Perhaps the greater concerns are the economic consequences. This is also a misconception. An economy without fossil fuel is a different economy, but not a worse economy. This new economy, based on highly efficient utilisation of zero-carbon energy, would generate a huge economic expansion to replace and upgrade infrastructure built through two centuries of reliance on fossil fuel. The jobs lost from killing the fossil fuel industry will be more than compensated for by new employment opportunities; but people are fearful of change, and want to keep what they have. Focusing on the negatives keeps us trapped and dependent on fossil fuel.

Focusing on the positives is the way to break out of the trap. When you examine the future beyond fossil fuel, the economic benefits of eliminating fossil fuel are huge. However, if you are a coal miner, own an inefficient poorly insulated house, or expect to continue flying in the current generation of thirsty jets, your personal circumstances may lead you to view the transformation as negative. It all boils down to attitude. A slow deliberate evaluation exposes the negatives as short-term and therefore of little lasting concern, but highly relevant to those with vested interests in the current fossil fuel economic model. Such slow deliberate evaluation shows that, for society as a whole, the transformation to a world beyond fossil fuel dependency is overwhelmingly positive.

Perhaps it is our lack of imagination and inability to visualize a society and economy without fossil fuel that holds us back. If we could see how communities rescued from fossil fuel dependency are clean, vibrant and revitalized, we would want to live in such places.

There is less choice over responding to the consequences of climate change. Some of these consequences are already with us, such as: more severe storms, heatwaves that are hotter and droughts that are more sustained. This is just the beginning; over the coming decades changes which are already locked into the climate system will become apparent. Despite the best efforts of climate scientists, we cannot know the exact outcome until it unfolds.

In responding to the consequences of climate change, we will have to do our best to carry on. For regions at risk of reduced rainfall, these will not immediately change to deserts. It will be possible to carry on using groundwater to irrigate crops for many years after rainfall becomes rare. As levels of groundwater drop, boreholes will need to be deeper, until even the deepest boreholes run dry. Eventually, agriculture would have to depart and allow the desert to take over.

In responding to sea level rise, the first response in coastal regions will be to consider sea defences. The mean global sea level rose by 20 centimetres through the twentieth century; and it will continue to rise as the water which has been held as ice in glaciers and ice sheets on land melts and flows into the ocean. London's prime defence against the sea is the Thames Barrier, which opened in 1982. It was designed to withstand tidal surges but not designed to cope with rising sea levels. The UK Environment

Agency have assessed that with enhancements it will continue to protect London until 2070[9] but it is likely that a new structure will be needed before the end of the century. The bigger long-term decision for London is when to abandon efforts to hold back the tide, stop investing in defences and accept the inevitability of losing much of the city to the sea. Such momentous decisions do not yet make it onto the radar, but before long will have to be considered. The measures required to respond to the consequences of climate change will become ever more extreme and we will have no choice but to go with it. All because we have been so resistant, over so many years, to make the prime decision to give up on fossil fuel.

## Adaption to climate change

It is worth considering specific examples of adaptation to draw out the potential scale of the challenge and the need to reconfigure the infrastructure on which society depends.

At the most extreme end of the scale is the Maldives. The average height of land in the Maldives is less than 1 metre above sea level. It would only be with an immediate and total ban on fossil fuel that there could be any chance that this tiny nation of low-lying islands could be saved. It would be theoretically feasible, in engineering terms, to build platforms and structures above the advancing sea, but in any realistic sense the islands of the Maldives will no longer exist sometime later this century. The nation of people who are the Maldivians need to purchase land to make a future elsewhere. The realistic policy for the Maldivian government, after trying and failing to persuade the

---

9   Environment Agency (2012), 'Thames Estuary 2100, TE2100 Plan.' Available online: www.gov.uk/government/publications/thames-estuary-2100-te2100 [accessed 26 July 2020].

world to rein back on emissions, is to shame the world into paying for their new home. The same applies to a number of low-lying island states which in 1990 formed the Alliance of Small Island States (AOSIS) to consolidate their concerns and increase their influence. This intergovernmental organization has been an important voice in the climate negotiations but sadly their best chance now seems to be to fight for cash handouts as compensation for damage, rather than relying on a robust agreement to limit climate change.

It is not all doom and gloom. For the UK there will be winners and losers. Winners will be English winemakers. The climate of Southern England is now warmer, and the summers drier, which is good for growing high quality grapes. English wines used to be the brunt of many jokes and only served when the host wanted to introduce a talking point. This is no longer the case; English sparkling wine has started to win competitions based on blind tastings. English wines have attained sufficient quality to impress the experts and are escaping the shadow of their old reputation. English wine is served, not because it is a novelty, but because it is good. Climate losers in the UK include future generations of Londoners living in low-lying boroughs such as Southwark, Whitechapel, and Westminster. There are also iconic landmarks lined up for inundation that will have to be evacuated sometime this century, or next – including Westminster Abbey, Canary Wharf and the Houses of Parliament. For this historic building, a surreal debate is taking place over a multi-billion-pound refurbishment. The Palace of Westminster has stood for nearly a thousand years since Westminster Hall was completed in 1099. Spending billions of pounds on refurbishment would make sense from the perspective of investing for the next thousand years. The reality is that such refurbishment has relatively short-term payback. The long-term investment should be going into a location in a new capital city designated on higher ground. It is ironic that the building where discussions over UK climate policy take place may become a prime example of wasted investment because of failure to tackle climate change effectively.

The US coastal state of Florida is particularly at risk. For example, most of Miami is barely 1.5 metres above sea level. The most desirable property is located on the waterside or beside the beach. This is also the property which is most vulnerable. There are already areas which are uninsurable because the risk of flooding is now so high. Rising sea levels and increasing severity of storms will hit Florida hard. Residents will have a number of choices. The first is to sell up now – whilst there are still people willing to buy (who do not understand the consequences of climate change) – so the risks can be passed on and the capital reinvested in houses on higher ground. Another approach for elderly residents is to decide that they need their beachfront home for only a decade or so, and after that, so what? People who decide to stay will find insurance becomes increasingly unaffordable so should expect to live in houses without insurance for their final years. The downturn in Florida property could be swift, as sentiment can change quickly. Maybe it will take a particularly intense storm causing a tidal surge and extreme flooding to drive home the message. Describing the situation in words and explaining the risks gains little traction. Newsreel footage and social media content showing flooded inundated areas and trashed properties may be more effective. Such accessible news could force people to face the reality of the risks of living in Florida as sea levels continue to creep higher. It might be slow, but year-on-year each centimetre of sea level rise is adding to the risk. This ratcheting up of danger is now inescapable. As uninsurable homes fail to find buyers, prices will fall and some home owners move into negative equity; foreclosures will follow. A few people, who have always fancied living in the best parts of Florida, could take a punt that they can live the high life for long enough to capture happy memories before their number comes up in the climate change lottery and they are forced to move. It seems impossible at this moment that Florida could lose its sheen as a favourite holiday and retirement destination, but it is hard to see how the Florida authorities could manage a transition without huge resources to demolish low-lying property to rebuild in line with expected higher sea levels.

Bangladesh is another example where the consequences of climate change will hit the country hard with knock-on consequences for the world. Two-thirds of Bangladesh is less than 5 metres above sea level. As sea levels rise, millions of people living in its low-lying coastal regions will be displaced. Sea defences are not an option. To protect such a massive river delta would require resources completely beyond the capability of any country, let alone one as poor as Bangladesh. As the land is abandoned and reclaimed by the sea, the consequence will be a huge wave of climate refugees. This flood of people will be way beyond what Bangladesh can cope with internally. So, as Bangladesh is swamped by the advancing sea, the world will be swamped with refugees looking for a new home. Rich countries who have relied on fossil fuel for so long – despite clear evidence of the consequences for the global climate – will have a moral, and perhaps a legal, responsibility to open their borders. If they refuse, they risk incubating a huge terrorist backlash from people disenfranchised by fossil-fuel-induced climate change.

Throughout the world each country will have its adaptation challenges. For some countries the consequences could be severe. For others, it could be neutral or even positive. Finland is an interesting example, perched on the edge of the Arctic where it might be expected that a warmer climate could be welcomed, particularly for agriculture. I had a discussion with a Finnish government official who explained that they understood that climate change is a global problem which will have global consequences, and that Finland will play its part to prevent it. This is an example of the sort of statesmanship the world needs to work together. Adaptation will become increasingly necessary but the only real solution, whatever the county's particular circumstances, is to cooperate to close down the fossil fuel economy.

# Mitigation of climate change

Instead of working towards closing down the fossil fuel economy, global efforts so far have been focused on the 'mitigation' of climate change. This is the overarching term used by the UNFCCC for a range of actions to reduce the concentration of carbon dioxide in the atmosphere.

> ... *the key for the solution to the climate change problem rests in decreasing the amount of emissions released into the atmosphere and in reducing the current concentration of carbon dioxide ($CO_2$) by enhancing sinks (e.g. increasing the area of forests). Efforts to reduce emissions and enhance sinks are referred to as 'mitigation'.*[10]

Mitigation is a weak term inappropriate to the scale and importance of the task. According to the Oxford English Dictionary it means:

> *The action of reducing the severity, seriousness, or painfulness of something.*

---

10  UNFCC (2020), 'Introduction to Mitigation' Available online: https://unfccc.int/topics/mitigation/the-big-picture/introduction-to-mitigation  [accessed 17 March 2020]

The use of the word 'mitigation' implies that it might be relatively easy to reduce overall carbon dioxide emissions through a range of relatively painless measures. It would be better to speak plainly and state the truth that we need to stop burning fossil fuel and close down the fossil fuel economy. But that sounds like hard work, so we use weasel words like 'mitigation' to hide the scale of the challenge. It should be no surprise that climate conferences spanning over two decades have produced a lot of hot air but precisely nothing in terms of real action. I will propose in the next chapter that the time has come to stop hiding the scale of the challenge and come up with a more effective response.

## Closing down fossil fuel

The stark reality is that we cannot stop climate change, but we can stop making the situation worse. This will mean stopping releasing fossil carbon into the atmosphere. To express the challenge in precise words, the challenge is to close down fossil fuel consumption. Let me repeat the challenge: 'to close down fossil fuel consumption'. This is clear, precise and exactly matches the situation we face. Many people recoil at such clarity, fearing that it would be far too difficult. It is easier to pretend that mitigation measures and voluntary agreements to emit less carbon dioxide could do the trick. Such convenient bluster does not convince me. In my view, the activities of the UNFCCC have become an excuse for sustained inaction. Governments can point to the UNFCCC and claim that the problem is being addressed instead of getting on with doing something. Waiting for the UNFCCC to find a solution is going to be a very long wait. Better to accept that the UNFCCC process cannot, and will not, deliver the solution. This should open up space to get on with establishing a process that can.

So, the only effective way to address the climate change challenge – to an extent that will make a real difference – is to close down fossil fuel consumption. The fact that we have not already done so, when the evidence is so clear, is a mistake on a scale not seen before in human history. It is probably not an exaggeration that climate challenge is unprecedented, and the required response is also unprecedented. Those of us who live in Britain could sit tight in the relative safety of our little island, to wait to see what unfolds. This is an option, but it would be a selfish choice which conflicts with UK aspirations to be a leader on the world stage. Leadership in the twenty-first century requires commitment to be on the winning side in facing up to climate change. Countries which fail to support collective action should become pariah nations, in the same way that countries which host international terrorism are excluded from the benefits of the world community.

If we don't face up to climate change, the challenges this will bring to society, and the increased security risk, will be substantial. People in rich countries might think that they can close their borders and use their purchasing power to outbid poorer countries to keep access to essential supplies such as food. But in a world where more and more places are at risk of economic and social collapse, ungoverned spaces could multiply and a series of conflicts undermine society's ability to cope.

## Making the right choice

One choice is to sit tight and do nothing. We live in a world which depends on fossil fuel, so we can choose to focus on defending our current lifestyle and protecting access to fossil fuel in order to carry on. We can continue to enjoy the throaty roar of the fossil-fuel-powered V8 engine SUV; we can continue to use cheap plastic items made from chemicals derived from oil; we can continue with our wasteful consumer lifestyles without interruption, without significant change. 'Experts' can be found who support such denial of the urgency to act. Nigel Lawson is an example of someone who holds views which can be used to justify inaction. He is no fool, and has held a number of senior UK government positions, including Chancellor of the Exchequer 1983–89. He is chairman of the think tank, the Global Warming Policy Foundation. He leads an argument which many people find reassuring that we should not be investing significant effort or resources in fighting climate change. His strongly expressed view has a following amongst a certain demographic and is deeply disturbing for displaying such deliberate wilful avoidance of the plain facts. As is often the case in such calculated behaviour, at the heart of the argument is an element of truth. The argument rests on the fact that every generation so far has been richer than the one before. The argument goes that it is not for us to invest now to deal with a problem which will manifest in the future. It makes no sense to spend our resources on fixing a problem which will impact on future generations. They will be richer than us so more able to afford the cost of fixing the problem. This somewhat warped logic supports the view of doing nothing and carrying on regardless. To my mind, this is irresponsible and clearly wrong.

The assumption that future generations will be richer is not a safe one. If our actions today are trashing the planet of tomorrow, then future generations will be poorer than us: poorer in environmental terms, poorer in agricultural terms, and poorer in every way as a society. A divergence is opening up between young people who want climate change action and are already seeing their economic prospects evaporate, and resistance from an older generation who own the bulk of the assets. The self-serving idea, expressed by those with wealth and power, that action can wait should be expunged from the climate debate.

I like to describe the sort of people who adopt the Nigel Lawson view as 'clever idiots'. I find it easier to accept people who don't understand the risk and close their minds to taking action on their own initiative. They will go with the flow, leaving people who do understand the need to act to lead the way. They are normal members of society. However, preaching inaction based on spurious logic is a level of incompetence of which they should be ashamed.

The clever idiots can be engaging and reassuring in what they say. It is so much easier to be told to sit tight and do nothing. However, it is getting harder and harder to reconcile this convenient message with common sense. As the cult of people who deny the need for action shrinks, and their message looks increasingly implausible, it is human nature to defect and want to join the tribe who are winning the argument. People will want to align with forward-thinking people of reason, with whom we can feel a sense of identity, rather than remain under the influence of an oddball cult which is increasingly marginalized. You can choose to do nothing until forced to act; that is your decision. That would be selfish but logical. To deny climate change, and pretend that it is not a threat, is plain stupid. In my experience, people don't mind being branded selfish; perhaps this is a by-product of modern living. However, no one wants to be branded stupid. There is therefore a good chance for the debate over climate policy to leave the clever idiots out in the cold.

Making the right choice requires intelligent thought and the resolve to follow through. This is a situation where ignorance is certainly not bliss, or not for long. As reality bites, denial of the dangers of climate change is exposed as indefensible. Avoidance of taking action is selfish. Signing up to superficial action is foolish.

Continuing to support discussions which have gone nowhere for decades is naïve. The stalemate which grips society is selfish, foolish and naïve. Making the right choice requires determined people willing to break out and accept reality, take responsibility, and push ahead with action. There are viable choices. We just need the courage and resolve to get on with it.

# Framing the decision

There are distortions swirling around the climate debate, and unfortunately succumbing to half-truths is easy. Let us start by escaping misrepresentation from whatever source. This could be vested interests focused on continuing to reap profits, knowing that decisive action will put them out of business. It could be politicians saying whatever is convenient to get elected, knowing that they are peddling false hopes and playing on people's prejudices and fears. It could be people who are so intently focused on self, and defence of what they currently have, they have closed their minds to logical analysis. All these sources of misinformation are part of a tsunami of comment and discussion which swamp the climate debate. Seeing through it all is not easy, and requires effort. One way is to let other people make the effort to do the analysis and then choose carefully which experts you can really trust.

As well as choosing who to trust, you should choose who to ignore. Steer clear of the populist politicians but also beware the clever idiots who, because they believe their own rhetoric, can sound very convincing. Particularly steer clear of anyone whose livelihood derives from continuation of the fossil fuel economy. The people who deny the need to act, and who will be poorer as a result of closing down fossil fuel, will get what they deserve.

The world needs open-minded people who are willing to accept that there must be change, and that the changes will be both positive and negative. I believe it is entirely reasonable for such people to get ahead by anticipating the coming changes and reconfiguring their affairs to reduce the risk to their lives. It is even reasonable, I believe, to plan to profit from such foresight to reap the rewards as the changes unfold. It is sensible that our first response is to protect ourselves, our families, communities and nation. Individuals should move away from low-lying coastal locations, and do so before property prices collapse. When it becomes common knowledge that this makes sense, it will be too late. You need to move before the crowd cotton on. When it comes to property you own, after ensuring it won't be at risk of flooding, make sure it is well-insulated and energy efficient. If you are a concerned person you may take pride in reducing your carbon footprint, but the real reason is because energy prices in the new economy are likely to be higher and you may have to run your home using only the renewable energy available locally. In places with consistent sunshine energy might be plentiful (and may even be cheap) but in other places renewable energy might be at a premium and it could be a struggle to reduce energy consumption to match the energy available. This is a complete reversal of the current energy model where power stations are built to match demand. It will take time for the new economic reality to be accepted and understood. Protecting ourselves requires deep forward thinking within the context of curing our addiction to fossil fuel. Yes, of course, we will suffer 'cold turkey' but we end up in a much better place. For the issues discussed in this paragraph, the outcome is living on high ground free from flooding in quality houses that are cheap to run. There is nothing negative about the outcome, except that you need to make an effort to make it so.

At the national level, there should be plans to deal robustly with climate refugees. Some countries will feel a sense of responsibility and offer to accept climate refugees, but they will need strong and effective arrangements to assimilate them. Otherwise they might become caught up in climate-related terrorism. Such terrorism is likely to be a feature of a world where some places are hit hard by climate change. Poor countries with the least historic responsibility for carbon dioxide emissions could become very angry places and terrorist breeding grounds. Other countries will take a tougher line and implement tight immigration rules and attempt to strengthen their borders. The main point is that nations need to think well in advance and have plans to prevent social unrest despite a flood of desperate people doing whatever they can to escape the climate chaos as whole communities are forced to move.

Food security will rise up the agenda of national concerns. Currently rich countries can buy what they need, outbidding those less able to pay, so that hunger is confined to poorer countries. In a world where agricultural capacity is under threat from climate change, and countries start to reserve local produce for local consumption, countries which are now currently net importers of food (such as the UK) will need robust arrangements to secure reliable supplies.

It seems unlikely that there will be robust action over climate change any time soon. As a species with one habitat, planet Earth, we should be adopting a risk-averse approach rather than meddling with our life support systems. Unfortunately, we seem unable to bring ourselves under control. This means we must prepare for the long game expecting climate change to get significantly worse. We can only be sure of being safe in a lawless world, awash with climate refugees, by making preparations. At the macro level, the safest countries will have good security, high-quality infrastructure, and considerable self-sufficiency. At a local level, particularly in more vulnerable countries, it might be safer to live in gated communities to be secure as lawlessness spreads. Ideally such communities would be along the lines of eco-communities, largely self-sufficient in many respects. This might include energy self-sufficiency from solar and wind combined with community projects to grow food in sheltered areas and in greenhouses to have crops throughout much of the year. Such eco-communities may seem like an indulgence for Greens but, in a world in turmoil, they could be the refuges we need to remain safe. An eco-community may not be an indulgence but a necessity in a world in flux.

I have focused here on our first response to protect ourselves, our families, communities and nation. Our personal affairs need to be firmly rooted within a community, with secure access to resources, and in a country with good security and robust policy to thrive in a changing world. From such a position of strength, we might reflect on why we need to consider such measures to secure our future. We might then wonder whether halting further pressure on the climate at a much earlier stage would have been worth considerable effort and investment. From such reflection, it soon becomes apparent that moving quickly to address the cause of climate change is of course the right response, the best response, and the only response that makes any real sense. For people at the very edge of survival, the focus is on getting by from day to day. For people with resources and security, the focus should be on going forward as fast as possible to cure fossil fuel addiction.

## Backtrack on fossil fuel dependency

One way to make a difference with regard to climate change is to backtrack out of fossil fuel dependency. It would not be easy, but at least it would be simple to define. The alternative to backtracking is to leap forward, which again will not be easy and much harder to define. I do not recommend backtracking but it is worth examining for two prime reasons. First, it helps to explain why people are so resistant to addressing the climate change challenge. Backtracking has negative connotations and does not look attractive. Second, by defining the backtrack option it provides a canvas on which to paint leap-forward options. We must decide whether to backtrack or leap forward, as we must do one or the other. To continue on the current path is suicidal and not a credible way out of the predicament we have made for ourselves. We have to make a choice. Let us consider which is preferable, starting with backtracking.

We can get a flavour for backtracking by looking at three core components of our way of life. These are: agriculture, buildings, and transportation.

Take fossil fuel out of agriculture, and we can no longer rely on highly mechanized industrialized farming using large quantities of fertilizer derived from oil. The initial impact on agricultural output could be significant. Prices of agricultural produce would rise whilst farmers struggle to continue with much the same methods as now. Backtracking in agriculture looks like lower output and higher prices. It does not look attractive.

For buildings, take away fossil fuel and we will need to turn down the thermostat of the heating in winter and avoid using air conditioning in the summer. Currently, the UK electricity grid relies on 42% natural gas, 21% nuclear, 9% coal, 25% renewables, and 3% other fuels.[11] Cut out fossil fuel rapidly and there will be a huge deficit, power outages, rationing and escalating prices. We will find ourselves shivering through the winter and sweltering

---

11  Energy UK (2019) 'Electricity generation', Available online: www.energy-uk.org. uk/our-work/generation/electricity-generation.html [accessed 26 July 2020].

through the summer as energy is rationed and expensive, once fossil fuel is withdrawn. Backtracking on how we run our buildings does not look attractive.

Backtracking in transportation does not mean returning to using the horse and cart, but it would mean accepting restrictions on the transport options currently available. Flying is perhaps where the environmental debate is most intense. Restricting flying could be either through making flying expensive, which would be politically divisive, or through some form of rationing. Ideas have been floated that every person is allocated a certain number of flying miles which they can use or sell. Those who want to fly more than their allowance will need to buy more miles. There are all sorts of variations on how to implement rationing, but the outcome is restriction on flying and fewer affordable holiday choices. Backtracking on aviation does not look attractive.

For travelling by sea, backtracking could mean returning to sailing ships for both passengers and freight. This might become a romantic way to cruise but schedules dependent on the wind will be much less reliable. Freight arrival times will be uncertain and holidays will be of variable length dependent on weather conditions. The way schedules are planned would need to change to incorporate such uncertainty. Backtracking on sea transportation looks attractive from a nostalgic viewpoint but not in terms of reliability.

So, agriculture, buildings and transportation look like they would be hit hard by the rapid elimination of fossil fuel. When examined as changes to what we have now, backtracking does not look like an attractive future. It is no wonder that people baulk at action to deal with climate change, if this is the future it would bring on. It is necessary to think clearly and deeply to overcome our initial reluctance. Instead of considering what we might lose, we need to imagine what can be gained.

# Leap forward

Fortunately, the alternative to backtracking is to leap forward. It takes more planning, more forethought, and more investment, but it is entirely feasible. It does not require any magic new technologies yet to be invented. We have had, for many years, the means to leap forward but we do not see where to leap because our field of view does not extend beyond the current fossil fuel economy. We can see that backtracking is feasible, but unattractive. We don't have the imagination to see the landing place to want to take the risk of leaping forward. If we could see where the leap will take us, and understand that the landing, although potentially bumpy, is a better place, we might just find the courage to give it a go.

The required leap forward in agriculture comprises a different attitude to farming. Instead of monoculture at industrial scale, it requires a closely integrated system of crop and livestock production practices, appropriate to place and managed carefully to maximize yields by intense organic farming methods. Yields can be high, and the impact on the ecosystem reduced, as sustainable farming practices are deployed. In some respects, this is rediscovering the deep knowledge of traditional farming over the centuries but turbocharged by modern science. The downside – if it can be called a downside – is that this greater complexity requires much more human expertise and involvement. Modern farms have few workers and lots of expensive machinery; the new agriculture will generate jobs. These jobs need not be back-breaking manual labour but stewardship roles for the careful nurturing of natural systems and processes. This expansion in employment in agriculture will proceed in parallel with the reduction of employment in factories as they become increasingly automated. The resurgence of rural communities could see a huge increase in quality of life, as bland industrial agricultural landscapes make way for complex integrated biodiverse agricultural systems.

It is interesting that such 'new' farming practices may not be that new, but apply well-known principles from the past and use technology to reduce the hard, physical work. This resurgence in agriculture will include crops to recharge the soil in place of fertilizers and annual cropping to replace ploughing and sowing each season to reduce the need for mechanization (and conserve the soil). The final interesting observation of the leap forward in agriculture, is that yields can be as high as the industrial methods currently employed (although those with vested interests in the current industrial agriculture model may dispute this). Industrial farming survives now on the false premise that only industrialization (supported by fossil fuel) can provide enough food to feed a hungry world, justifying the high environmental impact. The leap forward in agriculture will take considerable effort to establish. Farmers will have to relearn techniques which work with nature rather than rely on the brute force of industrial practices. Those with a vested interest in the current industrialized agricultural system, selling equipment, chemicals and fertilizer derived from fossil fuels, will have to be faced down. It will be exceedingly difficult to win the argument, but the landing place for agriculture, after taking the leap forward, would be a welcome transformation of the countryside.

The potential to make radical improvements to buildings is inspiring, leading to much better places to live and work, but requires a complex mix of clever architectural design and advanced building technologies. Sustainable buildings are much harder to design than current standard designs where the building design comes first and then heating or cooling systems are bolted on as required. Buildings will need to be designed to suit the place and the prevailing climate. Methods to maximize solar gain will be deployed in buildings for northern climes, or minimized in hot climates. In hot countries the norm should be small windows, thick walls and plenty of shaded spaces. In cold countries, large south-facing windows will be used with features such as carefully positioned deciduous trees to provide shade only in the summer months. The fabric of buildings will be highly insulated and the windows will be triple glazed. Such windows are currently standard in Scandinavian countries. This is because the intense cold makes such high-quality features cost-effective within the fossil fuel economy. Our house in the UK is fitted with triple glazing and we have low heating bills to match. To get these fitted, we had to push back against 'expert' advice against such 'extravagance'. Roof structures in all geographies will be designed to maximize the use of solar panels and other renewable energy technologies. Such high-quality buildings will be more expensive to build. Owners will either have to pay more, and borrow more, or accept smaller buildings of higher quality.

The evidence that the leap forward in buildings is feasible is there for all to see – if we bother to look. An example is the BedZED village in South London, designed by the architect Bill Dunster and completed in 2002. By careful design and investing heavily in insulation and other building components it was possible, in the benign climate of the South East of England, to eliminate the need for a heating system – thereby eliminating heating bills. In the current fossil fuel economy such heavy capital investment is not judged to be cost-effective because energy is cheap. It is cheaper to run a gas boiler than to build a much better building. The barrier to leaping forward is not technology but attitudes and economics. The primary barrier is the economics of the fossil fuel economy which favours cheaper lower-quality buildings. Another barrier is that house buyers don't understand what is possible: they don't demand such quality, they don't expect it to be provided and, certainly in the UK, they don't want a house without a conventional heating system. The final barrier is that many people in the building industry are lazy, making profits using tried and tested methods and not willing to take the risk of leaping forward. It would be wrong to accuse the whole industry of being lazy, but it is hard for those who understand the required transition to ply their trade when they encounter such entrenched resistance.

For the leap forward to take place in housebuilding, an entire industry will have to learn how to build efficient high-quality houses – and for all of us to demand such houses. There will be multiple points of resistance, arising from laziness and a lack of imagination, supported by an outdated economic model. This might prove hard to overcome, but the consequences will be that we live in much better houses.

The transformation in transportation has already begun. Electric cars are making inroads, and showing that they are as good or better than fossil-fuel-powered cars. The future for driving looks bright as advances in batteries improve range, and particularly when we move beyond batteries to adopt hydrogen fuel cells to power cars. Current petrol and diesel cars have a direct lineage back to the early mass-produced cars of a century ago. Interestingly, some of the very first vehicles were electric-powered with batteries. The technology is simpler than the internal combustion engine but fossil fuel was easily available and became established as the fuel of choice. The mass production of electric cars is long overdue. The car industry is now entering a difficult transition. It is being forced to make the leap by the troublesome Elon Musk who has launched the Tesla to shake up the car industry. Industry executives have had to wake up and spring into action. We will see this happening more and more; as people and industries with a vested interest in the fossil fuel economy defend, defend, and defend again, until finally a combination of upstart competitors and tighter legislation threaten to put them out of business.

The leap into the future for aviation is the most exhilarating transition in the transportation sector – and the most difficult to orchestrate. Currently, the workhorse of the sky is the fast jet. As a passenger, whether you fly First, Business, or Economy class, it is a seat on a fast jet. Flying fast needs a lot of energy; so fast jets are thirsty. The current aviation model therefore burns a lot of fuel. The leap forward is to accept that affordable flying will be slower. Examples of highly efficient relatively slow air vehicles exist but there are no plans to develop them into mass market aircraft. The reason is simple; with fuel so cheap, they are not commercially viable. The aerospace industry has the expertise, the technology, and the wherewithal to design and build slow efficient air vehicles. The block once again is the fossil fuel economy, but this time it is backed up by outdated and inflexible international aviation regulations which in effect prevent the taxation of aviation fuel for international flights. Until this changes, aircraft manufacturers and airlines cannot build a business case for highly efficient flying, particularly if this means persuading passengers to accept flying more slowly. Again, we find economic barriers and immovable attitudes come together to make leaping forward seem impossible. To overcome the economic barrier, we need to tax aviation fuel. To overcome the attitude blockage, we need to deploy relatively slow and efficient air vehicles which are safe, comfortable and spacious so that people can be persuaded to switch from fast jets. For those on a budget, it will become the norm to trade a speed reduction for an affordable ticket. We have all the technology and know-how to transform the model of aviation; all that is lacking is the determination to get on with it.

We need to tell our politicians that we will accept short-term disruption to be able to have sustainable flying options in the future. There is a vibrant future for twenty-first-century aviation waiting to be unlocked. The transformation could be quicker than anyone currently imagines, as soon as we back off from defending the aviation services we currently enjoy.

My analysis of climate change and aviation was completed before the COVID-19 crisis. During the final edit of the manuscript, I was in virus lockdown sitting in the garden under a clear blue sky devoid of vapour trails. It gave me the chance to reflect that the transformation may have already begun. The advances needed in aviation will require considerable investment, and almost certainly lead to a wave of bankruptcies in the current industry. My analysis concluded that this would not be initiated until there is coordinated action to rewrite international aviation agreements. I did not factor in COVID-19. Although regrettable for those who have lost jobs, the pandemic has provided a golden opportunity to restructure the industry.

Transforming aviation requires abandonment of the current industry and shifting onto a different development path which offers passengers different options using radically different air vehicles. Leaping forward in aviation is so difficult (and so potentially inspiring) that I have examined this sector in detail and found that it can be done. I relish the transformational leap forward in aviation, because from my engineering background I know it is possible and I can see that the landing place is so much better. People in a hurry, who can afford high ticket prices, will still travel by fast jet. Many more people will travel on a new breed of relatively slow air vehicle. Even time-poor business executives might find it better to travel slower to work and sleep in comfort. Economy passengers should welcome more space instead of being crammed into tight seats with little leg room.

It is serendipitous that these new spacious air vehicles, as well as reducing environmental impact, can also resist the spread of disease by having the space to allow social distancing, compared with the tight confines of fast jets.

In Chapter 10, I explain in more detail how aviation can be transformed. This really could be the sector in which to test our resolve and demonstrate that we can face up to climate change.[12]

For transportation by sea, the leap forward requires, once again, a new economic model facilitated by rewriting the rules to tax fuel used by shipping. Like aviation, the shipping industry does not currently pay tax on fuel, locking the industry into dependency on cheap fuel. Break the lock by taxing fuel and you create the incentive for ships powered by wind. These would not be the sailing ships of the past, but use state-of-the-art technology to capture energy from the wind and be able to operate in all conditions. When the wind drops, massive fold-out solar arrays can be deployed feeding battery storage and electric engines driving the propellers. Naval architects and engineers are going to have a whale of a time when they are invited to design efficient, clean ships to transform twenty-first-century sea transportation. These experts have the technology, the ability and the design skills, but the investors who will need to fund it don't have an economic model which can pay for it. I keep banging on about abandoning the fossil fuel economy allowing a new economy to rise out of the ruins of the old, because the opportunities are so inspiring. Transportation can be so much better, and I am confident will be so much better, but only when we all club together to face down the vested interests currently lined up in opposition.

---

12  McManners, P.J. (2012), *Fly and be Dammed: What now for aviation and climate change?* UK: ZED Books.

So, the leaps forward in agriculture, buildings, and transportation will require considerable investment, generating significant numbers of additional jobs, and be incredibly complex to orchestrate. It will be worth the effort because the landing place is so good. A world of high-quality buildings in tune with place; agriculture more localized and kinder on the environment delivering high-quality food; and transportation without the filth and the fumes of fossil fuel. What is there not to like about leaping forward?

Compared with backtracking, leaping forward is much more difficult for governments to arrange without losing popular support, and much harder for corporations to navigate without customers demanding it. If we could all embrace my vision of a world beyond fossil fuel dependency, and accept that the economy will be different, it would be so much easier. Governments would then have our support to change the economic model and corporations would then have the clear commercial incentive to push ahead.

## Conclusion

The prime decision is surprisingly clear-cut and simple. The only response to climate change which makes sense is a bold decision to close down fossil fuel consumption. Pretending that there is an alternative and easier way forward helps no one. The transition period will be hugely challenging – but let's get to it because the outcome is clearly good. We are held back by fear and laziness. So many organizations are dependent on the fossil fuel economy, and so many current jobs tied to it, that we fail to see that new organizations and new jobs will be generated to replace those which will be lost. There are opportunities with huge complexity and scale, which are there to be grasped. We should leap forward with confidence, even though we cannot exactly map the landing. The transition period will be hugely disrupting, during which those with foresight and drive will do well, and those organizations and businesses which resist can expect to suffer. All it takes is enough of us to insist that: yes, we can; yes, we should; and yes, we expect our leaders to get on with it – or we will vote them out of office.

*The UNFCCC celebrates another meeting*

# CHAPTER 4

# FAILURE TO FACE UP TO CLIMATE CHANGE

The First World Climate Conference was held over 40 years ago in Geneva.[13] It was a scientific conference sponsored by the World Meteorological Organization (WMO). Working groups were set up to examine a range of topics including climate change. This was an appropriate start to understand what was happening. It is now over four decades later; we understand what is happening, and why, but little has been done in response. This is a colossal collective failure for which we are all responsible. No one should claim: 'Not me, guv'. We all know the situation and we are all guilty of doing next to nothing. Even those people who are making changes in their own lives – such as reducing their carbon footprint – are not doing enough to push politicians for change which applies to everyone. This chapter outlines the failure so far to face up to climate change.

The world's primary response to the challenge of climate change is the UN Climate Change programme. If you are already familiar with it, there will be little information here which is new to you. If your view is that action so far has been grossly inadequate, and unlikely to deliver a solution, then you probably don't need to read this chapter. If you believe that the UN Climate Change programme has the issue under control, and is leading

---

13  The First World Climate Conference was held on 12–23 February 1979 in Geneva and sponsored by the WMO. It was one of the first major international meetings on climate change.

the world towards climate security, you need to read on, because you need to wake up to reality.

# Good intentions – a story of failure

In 1992, the world came together at the Rio Earth Summit to discuss the pressing need to deal with environmental overload. It was clear that climate change was a clear and present threat to society and that action would be required. The United Nations Framework Convention on Climate Change (UNFCCC) was established and its first success was to agree the Kyoto Protocol, which came into force in 1997. The agreement committed developed countries to emission reduction targets but no targets were set for the developing world, effectively ignoring the huge growth of emissions taking place in India and China. The protocol also lacked any effective enforcement mechanism. This was far from perfect but over the space of five years could be regarded as a good beginning.

The UNFCCC continued to beaver away through the next decade, leading into what was billed at the time as a pivotal moment, at the United Nations Climate Change Conference in Copenhagen 2009. This closed without agreement.

The first sign of significant progress came with the agreement signed in Paris 2015:

> *This Agreement ... aims to strengthen the global response to the threat of climate change, ... including by ... holding the increase in the global average temperature to well below 2°c and ... making finance flows consistent with a pathway towards low greenhouse gas emissions and climate-resilient development.*[14]

14 UNFCCC (2015), Conference of the Parties, Twenty-first session, Paris, 30 Nov–11 Dec 2015, document dated 12 Dec 2015 [available online: https://UNFCCC.int/resource/docs/2015/cop21/eng/l09r01.pdf; accessed 21 Oct 2019]: Article 2.

The pledge to hold the increase in the global average temperature to well below 2°c was progress, but was the Paris Agreement more than hot air? At the heart of the Paris Agreement are Nationally Determined Contributions (NDCs). These are declarations by each country to reduce national emissions. This exercise was a showcase of good intent, but the sum total of NDCs fell well short of what was required – and no sanctions were agreed for non-compliance. This point in 2015, six years after the high hopes for Copenhagen, could indeed have been a pivotal moment if it had been acknowledged that the negotiations had, to all intents and purposes, failed. This would have opened up the possibility of a different approach and new direction. However, the UNFCCC was not going to be diverted quite so easily.

The next step forward on the current path took place three years later at Conference of Parties 24 (COP24) in Katowice, Poland 2018. The aim was to build on the 'success' in Paris to agree a rulebook to implement the Paris Agreement. In the lead up to the meeting, the need for action was reinforced by the Intergovernmental Panel on Climate Change (IPCC), the global body of the world's leading climate scientists, warning that allowing warming to reach 1.5°c above pre-industrial levels would have grave consequences. One person not heeding the warning was US President Donald Trump, who gave notice before the meeting that the United States would be withdrawing from the UNFCCC. To add to the sense of foreboding, the Polish President, Andrzej Duda declared at the outset that his country could not reasonably be expected to give up its 200 years' worth of coal reserves.[15] For the host country to set the tone with such words did not bode well for a successful outcome.

---

15 'COP24: Not all hot air', *The Economist*, 22 December 2018, 120–121.

The conference agreed an 'operational framework for ... tracking and evaluating efforts at the national and international level'.[16] In the UNFCCC Annual Report 2018, the section on the Katowice conference included this statement:

*... the package outlines how countries will report on their NDCs, the specific action they will take and how they can communicate their progress. ... While current pledges under the NDCs fall far short of where the international community needs to be to achieve its climate goals, by finalizing the Katowice climate package nations showed they are committed to increasing their ambition.*

The sum total of the fanfare of Paris 2015 and detailed negotiations of Katowice 2018 was 'current pledges under the NDCs fall far short of where the international community needs to be to achieve its climate goals'. Positive spin was applied by adding the words: 'nations showed they are committed to increasing their ambition'. The COP24 President, Michal Kurtyka, declared that the Katowice Package 'is something the world has been waiting for'. Really? This seems like a line from Samuel Beckett's play *Waiting for Godot*; a play in which nothing happens. We don't want statements that nations have increased their ambition (from a very low base). The world needs to face up to action.

Drafting a rulebook for an agreement without any enforceable provisions was always going to be a ridiculous exercise. And so it was. All this did was to give the impression of progress. As the Katowice conference drew to a close, once again the talks had achieved very little and the prospects going forward looked bleak. Brazil signalled its climate scepticism under President Jair Bolsonaro by withdrawing its offer to host the talks in 2019.

16  UNFCCC (2018), UN Climate Change Annual Report. https://unfccc.int/sites/default/files/resource/UN-Climate-Change-Annual-Report-2018.pdf [accessed 26 July 2020].

Despite such setbacks, the effort which went into attempts to present the talks as successful was like a media spinning machine running out of control. In the Christmas 2018 edition of *The Economist*, the Polish National Foundation paid to have a letter printed written by the COP24 President, Michal Kurtyka, which included the words:

> *Thanks to [the Katowice Rulebook], a great step will be taken towards realising the ambitions expressed in the Paris Agreement. Ambitions which will make our children look back on our legacy and recognise that their parents made the right decisions at an important, historical moment.*

Claiming success was disingenuous. First, the ambitions expressed in the Paris agreement are not ambitious at all. Second, the Katowice Rulebook has little substance and no means of enforcement. Third, it will be more likely that the delegates' children will look back on the wasted opportunity and want to disown their parents. The only point that is correct in this state-ment is that this was a historical moment. 2018 was the moment when the UNFCCC confirmed that it is incapable of achieving real progress in dealing with the challenge of climate change. Once people see past the spin and accept the UNFCCC for the distraction it is, it will indeed be an historic moment opening up the space for real measures to face up to climate change.

I accept that it is unfair to single out Poland, as the whole world community is complicit. In defending coal, Andrzej Duda was seeking to protect Poland's economy. In spinning COP24 as a success, Michal Kurtyka was trying to be positive. In a sense they were both doing the jobs allocated to them to the best of their ability. Neither man is personally guilty but they are part of the world's collective failure.

# UN Climate Change Programme

Let us summarize the UN Climate Change programme. It consists of the UNFCCC, the Kyoto Protocol and the Paris Agreement of 2015. This section is highly critical and I risk offending the many good people who are trying to make a success of a flawed process. However, this is not the time to pull any punches. Failures need to be addressed rather than glossed over.

> **The United Nations Framework Convention on Climate Change (UNFCCC)**
>
> The UNFCCC web site describes its work as:[17]
>
> *In 1992, countries joined an international treaty, the United Nations Framework Convention on Climate Change, as a framework for international cooperation to combat climate change by limiting average global temperature increases and the resulting climate change, and coping with impacts that were, by then, inevitable.*
>
> *By 1995, countries launched negotiations to strengthen the global response to climate change, and, two years later, adopted the Kyoto Protocol. The Kyoto Protocol legally binds developed country Parties to emission reduction targets. The Protocol's first commitment period started in 2008 and ended in 2012. The second commitment period began on 1 January 2013 and will end in 2020.*
>
> There are now 197 Parties to the Convention and 192 Parties to the Kyoto Protocol.

The numbers relating to the UN Climate Change programme are impressive. It has been running for approaching three decades and there have been 25 sessions of the Conference

---

17 UNFCCC 2020, The History of the Convention [available online: https://UNFCCC.int/process/the-convention/history-of-the-convention; accessed 17 March 2020].

of Parties (COP). At COP24 in Katowice, Poland, there were 22,924 participants, from all corners of the world including 29 heads of state and 82 government. ministers.[18] Through the life of the UN Climate Change programme, thousands of tonnes of paper have been used to print reports, and many millions of air miles flown. This has been an annual jamboree on a huge scale.

Despite the huge expenditure incurred by thousands of people meeting each year, the UNFCCC has failed to broker an enforceable agreement to bring carbon dioxide emissions under control. The sum total of one of the longest and biggest bean-feasts[19] ever hosted by the UN, is absolutely nothing tangible.

Those who have been partying hard and continue to plan for the next big meet up will object strongly to my negative appraisal but we don't want good parties, conference dinners and warm words – the world needs action.

The primary achievement of the UNFCCC has been to deflect attention away from the real issues. It has generated a huge amount of analysis, assessments and (I write this with regret) hot air. The total impact is not simply a lack of progress but much worse. The UNFCCC gives the impression of progress, so letting governments off the hook. Governments can await agreement at the UNFCCC before acting. Every hard-nosed realist in politics can see this is true, but still the party rages on. The deliberations and draft reports are argued over and watered down to lessen their impact, and when finally published are spun to be presented in the best possible light. The UNFCCC has become an excuse for inaction, and if allowed to carry on in its present

---

18  UNFCCC (2018), UN Climate Change Annual Report. https://unfccc.int/sites/default/files/resource/UN-Climate-Change-Annual-Report-2018.pdf [accessed 26 July 2020].

19  Beanfeast – a celebratory party with plentiful food and drink, Oxford English Dictionary Online. [accessed 26 July 2020].

form could be an excuse for lavish conferences for decades into the future. Meanwhile climate change continues unabated.

# Madrid 2019 – COP25

As the manuscript for this book was being written, COP25 in Madrid was coming to a close.[20] I went through the documents and closing statements hoping to see evidence that would require me to amend my blistering attack on the process. There were a lot of words but the UN Secretary-General summed it up well:

*I am disappointed with the results of COP25. The international community lost an important opportunity to show increased ambition on mitigation, adaptation & finance to tackle the climate crisis.*

António Guterres, 15 December 2019.[21]

The UNFCCC spin machine struggled to squeeze out a positive message. Despite little substantive progress to report, the Executive Secretary of UN Climate Change managed to find these positive words to end her statement on the outcome of COP25:

*Together, with all sectors of the economy and societies at large, we must work tirelessly to address the greatest challenge of our generation.*

Patricia Espinosa, 19 December 2019[22]

20  Conference of the Parties, Twenty-fifth session, Madrid, 2–13 December 2019.

21  UN Climate Statement (2019), 'Statement by the UN Secretary-General Antonio Guterres on the Outcome of COP25' [available online: https://unfccc. int/news/statement-by-the-un-secretary-general-antonio-guterres-on-the-outcome-of-cop25; accessed 30 Dec 2019].

22  UN Climate Statement (2019), 'Statement by the Executive Secretary of UN Climate Change, Patricia Espinosa, on the Outcome of COP25' [available online: https://unfccc.int/news/statement-by-the-executive-secretary-of-un-climate-change-patricia-espinosa-on-the-outcome-of-cop25; accessed 30 Dec 2019].

The work of the UNFCCC seems to have fallen into a pattern that keeps the process alive. Like a weight-watchers' club which would close down if it was successful, there is little drive to get the task finished. Effort goes into drafting documents which outline unenforceable aspirations based on voluntary targets. These are written in such a way that they can fend off any real response until the next convention. Although I am scathing of the process, I don't blame the hardworking delegates. What we have is systemic failure. The UNFCCC may be beyond saving, but there is a UN organization which continues to have an important role.

The Intergovernmental Panel on Climate Change (IPCC) is the United Nations body for assessing the science related to climate change, established in 1988 jointly by the United Nations Environment Programme (UNEP) and WMO.[23] The IPCC is recognized as the most authoritative scientific and technical voice on the state of knowledge on climate change. The scientists on the IPCC have done good work to nail down the science so tight that it is now beyond doubt. If you meet and talk with these scientists you find that some of them are also disgruntled with the UNFCCC process. The UNFCCC is the only show in town, so most climate scientists keep their public criticism mute and carry on. They know that they are largely being ignored by politicians but hope that somehow their message will eventually get through. In reality, the good work of the IPCC will continue to be squandered within the UNFCCC, and will continue to be ignored. The UNFCCC process seems to have been designed to ensure that there will be no interruption to business-as-usual. It does not have the remit to challenge the dominance of the fossil fuel economy and therefore has no chance of succeeding.

---

23  United Nations Environment Programme (UNEP) and World Meteorological Organization (WMO).

The next stage will be more of the same; unless we find the wisdom, courage and determination to really make a difference.

# Limited success

The UN has had some success. It has brought together, in the IPCC, the world's most reputable climate scientists to gather and assess the research on climate change. The IPCC has provided some rock-solid insights. First, climate change is happening. Second, the prime cause of climate change is the burning of fossil fuel. Third, there will be serious consequences for the climate, sea levels, agriculture and the environment. Fourth, the precise consequences, their extent, and timescales cannot be predicted, so there will remain huge uncertainty which no amount of further analysis can eliminate. Climate change is real, it is caused by burning fossil fuel, the consequences will be serious, and there is huge uncertainty. These four conclusions are all that is needed to move on from analysis to start taking action.

The UNFCCC has made some progress by agreeing that the 'safe' limit to global warming is 1.5°C. The UNFCCC has recognized also that the lack of action so far means that this limit is not practical and almost certain to be breached. The UNFCCC has therefore stated that 2.0°C should be the absolute limit to avoid dangerous climate change. We are on firm ground with these numbers; they are not disputed and the UNFCCC has provided the forum to agree that these limits are appropriate.

In considering what to do next, I have reflected long and hard on why the UNFCCC is so ineffective. It is not just that negotiations led by the UN tend to be ponderous; this is inevitable when seeking agreement amongst disparate nations. It is not just arguments between the developed countries (blamed for their past emissions) and the developing countries (berated for their massive growth in emissions as they grow their economies). It is not just a lack of political support, although this is a key weakness. The problem with the UNFCCC is in its name. It is a convention on 'climate change'. This means it gets drawn into a wide debate about all manner of issues which are relevant to climate, but with little relevance to solving the core challenge. As an example, one of the issues which has been drawn into debate is the role of forests and their ability to capture carbon. Closely aligned with this is the reverse perspective, that chopping down forest releases carbon. This allows almost surreal arguments to surface such as blackmail threats to cut down forest if richer countries don't pay for its preservation. Some of the most toxic discussions revolve around carbon trading and carbon markets. The idea that these might be useful arose within the UNFCCC, and has gathered considerable momentum. It is argued that climate change can be ameliorated through trading carbon. It seems attractive on a superficial level, and has many supporters, so carbon trading is worthy of analysis. I will leave a closer examination until Chapter 9 where I show that, although carbon trading is a tempting way to appear to make progress, relying on a global carbon market would be seriously misguided.

The UNFCCC focus on climate change has had the benefit that there is now rock-solid knowledge of climate change and a strong grasp on the inherent uncertainty. However, continuing to focus on climate, as we discuss what to do, can completely miss the point. The action required is to close down fossil fuel consumption. Arguing over forests as carbon sinks and debating carbon markets is hugely distracting from this core objective. The retention of forests is important; and there is a potential role to trade carbon to smooth the transition within national economies under national control. However, drawing such issues into the global debate just muddies the water. We can end up in the exceedingly odd position of assuming that continuing to burn fossil fuel is fine provided some adjustments are made at the margins. The key challenge is climate change; the main barrier to making progress is the mistaken focus on climate. Rather than focus on climate, we need a laser-sharp focus on the solution, which is to stop the burning of fossil fuel.

# Conclusions

Responding to the challenge of climate change started well enough. Setting up the IPCC in 1988 was the right course of action to ensure the science of climate change was understood. Convening the Earth Summit four years later in 1992 was useful in getting international agreement that environmental overload was a serious issue which needed to be addressed, and that climate change was a particular concern. Setting up the UNFCCC followed quickly and was the next logical response to examine the issues and agree what to do about solving the climate change challenge. If momentum had been maintained, the challenge could now have been solved. Instead, the UNFCCC has become a block to progress.

The UNFCCC has been running so long, and failure to make progress so deeply entrenched, that the process of spinning success out of tiny gains has become an established game. The game continues on the basis that at least we are doing 'something'. Such optimism is not just ill-founded, but the situation is worse than most people realize. It is not simply misplaced optimism; the UNFCCC manages to convey the impression that action is in hand to deal with climate change. The consequence is that the UNFCCC is actually preventing real negotiations and real discussion by soaking up the limited resources available and providing a forum in which people believe they are being heard. It acts like a sponge soaking up points of view and publishing reams of paper. It is believed that because views and analysis have been recorded and documented by the UNFCCC, that they are serving a purpose. All that has happened is that they have been archived. Historians will be able to look back and examine in great detail how one of the greatest cons ever perpetuated in world affairs was played out. Rather than continue to fixate on the UNFCCC and its failings, I suggest we should move on. This is difficult because the world has put so much faith in the UNFCCC. The reality needs to dawn that the UNFCCC is not, and can never be, the forum to broker the solution.

The way the UNFCCC has lost its way is a disgrace. Most of the individuals directly involved are honourable people, but in my view the culture of failing to tackle the issues head-on is so deeply ingrained that the UNFCCC cannot be reformed. My proposal is that the IPCC retains the prime role as the most respected source of scientific advice on climate change and its consequences. A much smaller UNFCCC, protected from political interference and free to offer impartial expert opinion, may continue to be useful to keep the science under review. Resources need to be reallocated to actually solving the challenge, and rescuing the situation from gridlock, by focusing on the problems of fossil fuel.

*Giving up is hard to do*

# CHAPTER 5

# FOCUS ON FOSSIL FUEL

The prime cause of climate change is the burning of fossil fuel. So, the only way to deal effectively with climate change is to stop burning fossil fuel.

Over the decades that climate has been discussed, targets have been proposed and abandoned. Instead of decisive action, people and organizations are encouraged to measure their carbon footprint and take actions to reduce it. The great big carbon boots worn by North Americans, and medium-sized carbon shoes worn by Europeans, are compared with the minimalist carbon sandals of those living in Bangladesh. This focus on carbon footprints is a step forward, but in terms of identifying and agreeing pragmatic policy, it is not sufficient. The discussion needs to be more closely aligned to reality. It is correct to say that the cause of climate change is excess carbon emissions; but the cause of these emissions is the extraction and burning of fossil fuel. The solution is to be found in going direct to the root cause and in agreeing robust arrangements to ensure that fossil fuels remain locked safely away in the Earth's crust.

Focusing on fossil fuel is not just a better approach, it is the only realistic way forward. It should be borne in mind that the extraction of fossil fuel is very definitely something under our control – unlike climate change, which very definitely is not. Some fossil fuel can be accessed easily, such as where coal seams come to the surface. Such surface coal can be gathered up by hand and is thought to have been used by early humans to heat their caves. The Romans are known to have used coal in England in the second and third centuries (AD 100–200). In North America, the Hopi Indians used coal for cooking, heating and to bake pottery made from clay, long before European settlers arrived. These examples were at small-scale and so of little significance to the climate. The exploitation of fossil fuel today is a massive industry at huge scale, requiring high technology and large capital investments. This is particularly the case for oil and gas, which is found deep down at high pressure requiring cutting-edge engineering. This scale and complexity mean that the fossil fuel industry is challenging and is dominated by big corporations. This also means that it is relatively straightforward for governments to identify who is in control and to enact regulations to steer the future direction of the industry – supposing they choose to do so.

# The need for clarity

Looking at the climate challenge through the lens of fossil fuel provides much greater clarity and a crisp picture of possible ways forward. Debates about carbon are opaque and complex, leading to fudged policy which tends to support the status quo. So, it is not surprising that focusing on carbon has failed to deliver workable policy. Changing the approach to focus on fossil fuel opens the way to clear thinking and transformative solutions.

Focusing on fossil fuel goes direct to the core logic and exposes the dilemmas which cannot be ignored and have to be solved. Focusing on carbon is a relatively easy debate because it allows such difficult issues to be sidestepped. Making decisions about fossil fuel is tough because it forces policymakers to confront the real challenge. This is the way to make tangible progress. Policymakers should be persuaded to grasp the nettle, push past the huff and puff of the carbon debate, and focus on fossil fuel.

Let me pause here for a moment to concede (for the sake of readers who are enraged that I downplay the carbon debate) that the intellectual argument for focusing on carbon is sound. The positive case for carbon goes like this. First, we decide on the upper limit of carbon dioxide in the atmosphere which is judged to be safe. This is then set as the cap on total global emissions. If the countries of the world accept limits on carbon emissions which, when added together, are less than the cap, then hey presto, we have a solution. Translating this into reality is what the UNFCCC has been trying to do – and failing. You need to be incredibly naive to believe that all countries will sign up to sufficient carbon emissions reductions to hold atmospheric carbon dioxide concentrations within safe limits. Signing such an agreement would in any case be just the beginning; to be effective needs robust enforcement. This seems like a non-starter. I cannot envisage any circumstances in which debating global carbon dioxide emissions could lead to a workable pragmatic solution.

Instead of a narrow, intellectual and idealistic debate about carbon we need to face up to the stark reality of just how far society and the economy will have to shift, from where we are now to where we need to be. The fossil fuel debate, which is the focus of this book, is equally intellectually valid but its huge benefit is that it is grounded in the real world. This is the world dominated by national self-interest, where the fossil fuel industry is hugely influential and politicians tend to adopt the easy way forward, responding to people's immediate concerns rather than doing what is right for the long term.

Intellectual debates can only take us so far. We don't need grand conferences and further briefing papers. We need real policy which engages with the real world; and that means engaging with fossil fuel extraction, trading and consumption. In Chapter 9, I will consider how mechanisms such as carbon trading and a carbon tax can have a useful role to ease the transition within national economies, but any global solution with a chance of success will have to be about fossil fuel, rather than carbon. The case I make here to focus on fossil fuel is robust, logical and easily understood once you open your mind to the argument.

Not everyone reading this will initially be convinced by my argument, that we must focus on fossil fuel rather than carbon. If this simple change of tack can work, why have we not already taken this direction? The reason is, that so much discussion has swirled around factors other than fossil fuel that the idea persists that climate change can be dealt with through focusing on carbon dioxide emissions without confronting our addiction to fossil fuel. There are a number of myths which keep this fantasy alive. It is worthwhile examining some of the ideas coming out of the carbon debate to check whether there might be any traction or value in them. This examination of the alternatives is like the process an addict goes through before deciding that quitting and enduring 'cold turkey' is actually the only way. Discussion of other ways forward clarifies thinking and reinforces the understanding that closing down fossil fuel is the only way. Only when every other avenue is shown to be a dead end, will we accept that the difficult road is the only one which leads to the required destination.

## Four carbon myths

In my examination of alternative perspectives, I will explore four myths. First, that emitting carbon dioxide is not a problem because it is a normal component of the lifecycle of the planet. Second, that flatulating cattle are as much a problem as burning fossil fuel. Third, we are already making good progress with reducing carbon dioxide emissions, so let us continue in the same direction without significantly ramping up our efforts. Fourth, we will switch to other energy sources when supplies of fossil fuel run out, so the problem will be solved automatically without any special measures. If you are deeply engaged with the issues you will recognize that these perspectives are spurious at best, and in some cases plain wrong. Even so, the process of debunking them sharpens our understanding.

## MYTH 1

*Emitting carbon dioxide is not a problem because it is a normal component of the lifecycle of the planet.*

Carbon dioxide is part of the cycle of life, as schoolchildren learn at an early age. We, and other animals, breathe in oxygen and breathe out carbon dioxide. Fortunately, this is balanced by plants doing the reverse; capturing carbon dioxide out of the atmosphere and releasing oxygen in exchange. It is an entirely natural process which is vital to life, and must continue. For much of the planet's geological history, vegetation had the upper hand, stripping out carbon dioxide and incorporating the carbon into the structure of the plants and trees. Over millions of years vegetation has grown, died, and become locked into sedimentary rocks to become eventually coal, oil and gas. By the start of the human era, carbon dioxide in the atmosphere had stabilized at low levels with carbon in matched by carbon out. Plants would prefer that there was more; they could then grow more vigorously; and for farmers crop yields could be higher. Commercial greenhouse operators know this and, where they use gas to heat greenhouses, will often divert the carbon dioxide back into the greenhouse. It can be argued that increasing the concentration of carbon dioxide in the global atmosphere will benefit world agriculture; although it is doubtful whether increased yields would be greater than the loss of agricultural capacity in regions adversely effected by climate change.

So, it is absolutely true that carbon dioxide is a normal and safe component of our planet's systems. This cycle repeats, again and again, as plants grow (including crops and forest) and are then consumed by animals, burn or decay. Without the onset of industrialization, the carbon cycle would have been a closed loop, repeating indefinitely around a stable average level of carbon dioxide in the atmosphere.

Human activity, in extracting and burning fossil fuel, has upset this former stability. Carbon dioxide in the atmosphere has been monitored at an observatory located on top of Mauna Loa, on the Island of Hawaii since 1958. These are precise scientific records which show a trend line which continues inexorably higher, reaching a new record global average carbon dioxide concentration each year.

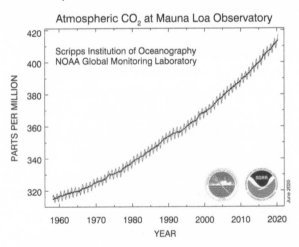

**Figure 5.1** Atmospheric $CO_2$ Levels at Mauna Loa Observatory[24]

---

24  Earth System Research Laboratories Global Monitoring Laboratory (2020). Source: NOAA Global Monitoring Laboratory [available online: https://www. esrl.noaa.gov/gmd/ccgg/trends; image generated 29 June 2020].

It is interesting that the line has a wavy profile, showing how $CO_2$ levels change with the seasons. It is like watching the planet breathe. The reason is that vegetation takes up more $CO_2$ in spring and summer. There are more forests in the northern hemisphere so global $CO_2$ levels are relatively lower when it is summer in the north. This cycle repeats each year as the lungs (forests) of the planet breathe in and out. This breath-ing cycle is entirely natural and normal; but the overall upward trend is not.

It would be wrong to blame the Earth's natural carbon cycle. This is not the problem; the problem is the industrial economy we have created which is liberating fossil carbon back into the carbon cycle. The carbon cycle is clean and normal; the extraction and burning of fossil fuel is not.

## MYTH 2

*Flatulating cattle are as much a problem as burning fossil fuel.*

A cow's fart contains methane which is a much more potent greenhouse gas than carbon dioxide. This is often mentioned as one cause of climate change alongside burning fossil fuel. This has an element of truth which is why this myth continues to have traction. There is plenty to complain about 'modern' methods of beef production. Industrialized beef production does indeed have a high environmental impact but it is a distraction to link these valid concerns with climate change. Grazing cattle in moderation is a normal part of a stable agricultural system. The intense meth-ods, where cattle are kept in feedlots and fed maize or soya beans, is not. Cutting down the Amazon rainforest, to clear land to grow soya beans, to export to feed cattle in the United States and else-where, is a dangerous industrialization of agriculture. Of course, beef production needs to be more sustainable – and such indus-trial processes outlawed – but cattle have always farted, and always will. Methane from cattle is a relatively small component of the natural carbon cycle. Flatulating cattle should not distract attention from the prime problem, which is the release of fossil carbon.

## MYTH 3

*We are already making good progress with reducing carbon dioxide emissions, so let us continue in the same direction without significantly ramping up our efforts.*

The growth of carbon dioxide emissions in the UK and across Europe slowed down between 2008 and 2019. This can be seen as progress, but these are still large carbon dioxide emissions. If you look closely at the reasons, the reduction is even less impressive. Part of this is because dirty industries have departed our shores as we have tightened environmental regulations. We still consume the products of these industries, but they are now made in other places such as China or elsewhere in the Far East. It is not a real reduction if all we have done is to ship the emissions to the other side of the planet. There is another even simpler explanation. The financial crisis of 2008 dented economic growth and put a brake on GDP. Perhaps lower emissions are simply a reflection of a weaker economy. As the world economy recovered the emissions again began to rise. For as long as we rely on fossil fuel, economic growth will lead to increased carbon dioxide emissions.

The claim that we are making good progress with reducing carbon dioxide emissions is false. The world has done very little to stop the rise in fossil carbon release, let alone make significant progress towards reducing it. Pretending we are on the right track, and making good progress, is simply an excuse not to get on with the real task of constraining fossil fuel extraction.

# MYTH 4

*We will switch to other energy sources when supplies of fossil fuel run out, so the problem will be solved automatically without any special measures.*

The myth that fossil fuel will run out and force us to find alternatives is fantastical. There are huge reserves of fossil fuel, in particular coal, which could keep the fossil fuel economy running for many more decades. There is a useful comparison that the Stone Age did not end because we ran out of stone. We developed better alternatives and moved on. It will be similar with fossil fuel. We will find the resolve to move to cleaner, better energy sources, leaving the remaining reserves of fossil fuel in the ground. The key question is how much damage will be inflicted on the planet before finding the resolve to cure fossil fuel addiction. The heroin addict will not be stopped because the land on which poppies are grown is used up, or because there are no longer farmers willing to make money from growing it. Heroin consumption is closed down because of laws which forbid the growing of opium poppies, combined with measures to wean addicts off the drug. Fossil fuel will not be eliminated because reserves run low or because there are no longer companies willing to profit from extracting it. Fossil fuel will be closed down because of a combination of laws which forbid the extraction of the dirtiest forms of fossil fuel combined with measures to wean society onto other energy sources.

Waiting for the coal, oil and gas to run out, whilst watching the planet burn, is not a viable policy option. This becomes even clearer in the next chapter when we look at the potential quantities of fossil fuel in the Earth's crust, and conclude that to safeguard the climate we can only afford to burn a fraction of the Earth's total fossil fuel reserves.

# Confronting fossil fuel addiction

Having exposed the myths, we know that the extraction of fossil fuel is not normal; the role of cattle is no more than a distraction; we have made very little progress so far; and fossil fuel will not run out. Having debunked the excuses, we can finally get around to solving the climate challenge. We will have to confront our fossil fuel addiction. This will require a dramatic reversal in our perception of the value of fossil fuel reserves.

It is heartening that there are already signs that the value of fossil fuel reserves is starting to be questioned. In 2010, Peabody Energy was the world's largest private coal corporation, valued at over US$1 billion. A significant part of this value was the coal reserves it owned. A number of sustainability experts, including myself, looked at the balance sheet and questioned the valuation. These considerable coal reserves would last for many decades. It was questioned whether society would allow the corporation to use these reserves to depletion. This message started to get through to investment analysts. Remember that these are hard-nosed investors, not dewy-eyed environmentalists. They understood the danger that society could persuade government to legislate to keep these coal reserves locked away underground. The share price started to tumble, culminating in bankruptcy in April 2016. This was Chapter 11 bankruptcy, part of commercial law in the United States which allows the company to continue trading whilst it seeks to repair its finances. I took this to be a healthy sign that a dose of reality was starting to force corporations to address a future beyond fossil fuel dependency.

So, the end is nigh for the US coal industry. Not so fast. A white knight came to rescue with the election of Donald Trump. He promised voters in the coal regions such as Alabama, Ohio, Pennsylvania, and Tennessee, that he would protect their coal industries. Perhaps 'dark knight' would be a better descriptive term for President Trump. The day after he won the US presidential election, shares of Peabody Energy surged more than 50% in over-the-counter trading. Peabody Energy emerged from bankruptcy in April 2017. The Chapter II provisions were an additional bonus to the corporation when, in October 2017, a judge ruled that Peabody Energy's bankruptcy protected it from global-warming lawsuits brought by California coastal communities against fossil fuel companies.[25] The action of Donald Trump has provided a delay, but anyone employed in coal will be out of a job eventually; and anyone who is heavily invested in coal will be bankrupt as soon as society faces up to climate change.

The death knell for coal is ringing out loud and clear. Whether the value of oil and gas reserves also plummets, is more complex to predict; but the value of fossil fuel reserves will come under scrutiny as the world gets serious about facing up to climate change.

Countries with considerable fossil fuel reserves, such as Saudi Arabia and Russia, derive wealth from their capacity to extract and sell oil and gas. As the world takes action to close down fossil fuel extraction, reserves which currently have value could become worthless. In Chapter 7, pragmatic policy will be considered for each fossil fuel to be treated differently. Countries with conventional oil and gas can enjoy one last blast of income from the industry's deathbed, but reserves of the dirtier fossil fuels will soon be worthless. If my proposals are embraced, and used to drive policy, this could be surprisingly rapid. As with

---

25 Randles, J. (2017), 'Judge Rules Peabody Energy Bankruptcy Blocks Global-Warming Lawsuits', Wall Street Journal, 25 October. [Available online: https://www.wsj.com/articles/judge-rules-peabody-energy-bankruptcy-blocks-global-warming-lawsuits-1508961807; accessed 23 October 2019].

Peabody Energy, the sentiment of investors can flip almost overnight when logical analysis is presented and accepted.

# Conclusion

The fossil fuel extraction industry is one of humankind's impressive achievements. Mining deep coal seams is difficult and can be dangerous. Drilling for conventional oil and gas is even more complex, particularly in offshore fields such as the Gulf of Mexico and the UK's North Sea. Fracking for gas or oil, to extract it from tight rocks such as oil shale, requires advance extraction techniques requiring special chemicals which need tight environmental controls. Fossil fuel extraction requires government permission and substantial investment. In the past, governments have actively encouraged fossil fuel exploitation because it is good for the economy (the fossil fuel economy). Change is on its way, starting with the coal mining industry, which is under intense pressure across the world, particularly in developed countries. Obtaining permission for new mines is getting harder and investors are steering clear. In the developing world, although the transition is slower, the World Bank has finally, and belatedly, backed away from funding coal mining and coal-fired power stations. The fossil fuel industry is not on its deathbed – yet. I believe the cleaner fossil fuels have one more profitable stage (outlined in Chapter 8) but this will be its last chance to reap profits before terminal decline.

When governments decide that controlling fossil fuel is required, it is within their power to do so. It is also within their power to conclude international agreements which limit the extraction of fossil fuel. Blanket regulations to close down all fossil fuel extraction would be simple in principle, but nearly impossible within the current political climate. What might be politically acceptable, and able to command popular support, is to regulate the dirtiest fossil fuels like coal, to in effect put the coal industry out of business. This is what is happening in the UK, where coal mining is all but extinct and the few remaining coal-fired power stations are obsolescent.

Dealing with climate change by focusing on fossil fuel is far more difficult than the current dialogue about carbon because it forces the debate to confront head-on the real challenge. By confronting the real challenge, it opens the way to a real solution. It is the difficult way forward; but in the end we will realize that it is the only way to face up to climate change.

What precisely needs to be done about fossil fuel depends on the total amount of $CO_2$ we can release before it gets dangerous. It is therefore the carbon equation to which I turn next.

*Would you ride this rollercoaster?*

# CHAPTER 6

# THE CARBON CALCULATION

## *One trillion tonnes $CO_2$ = 2 degrees*

The challenge of facing up to climate change is defined by the carbon equation. This tells us that the total amount of $CO_2$ we can release into the atmosphere to restrict global warming to no more than 2 degrees is one trillion tonnes. This 'allowable carbon budget' is much less than the known reserves of fossil fuel. The obvious implication is that consumption will have to cease long before fossil fuel runs out. On closer examination, it also becomes obvious that to make best use of the carbon budget, it should be used for the cleaner fuels such as gas, to get maximum energy for the carbon dioxide emitted. The overall result of the carbon calculation is the conclusion that we need a carbon budget based on the cleaner fossil fuels, with dirty fuels eliminated quickly without further delay. If this is obvious to you, you can dodge the maths of this chapter and go straight to the next chapter. If you want to know the detail, read on.

The basis of the calculation is a total budget for $CO_2$ capped at no more than one trillion tonnes. The key insight for policy is that this budget should be used only for the cleaner fossil fuels. I have simplified the calculation in a number of ways, and some of the detail can be debated, but the thrust of my argument is watertight.

The analysis presented here rests on a huge amount of number crunching which has already been carried out by the scientists of the IPCC. Their analysis is published in thick reports together with the supporting evidence. Their work has been detailed and thorough to take account of the complexity of the Earth's climate. There are a multitude of factors which the IPCC have considered, ranging from the fact that carbon dioxide is not the only greenhouse gas, to the fact that natural systems can absorb some of the carbon dioxide emitted. When you read IPCC reports you realize the depth and quality of the science. They bring as much precision as possible to something which is inherently unpredictable. The climate will alter in ways over which we have absolutely no control. The work of the IPCC is incredibly useful, not in the detail (about which there is uncertainty), but by nailing down the core logic. All I do here is to draw out the policy implications from this good work.

In this chapter, I will carry out some further number crunching, seeking to define the implications of the carbon equation. The maths I present here is simple. This is partly because I am focusing on the key factors and the big numbers, but also because I can lean on the excellent work of others. Where a reputable scientist or trustworthy source has published solid analysis, I use it. I don't claim any credit for the scientific analysis. All I have done is to expose what the maths tells us when focusing on what really matters. I have not allowed myself to get caught in the trap of making early assumptions about what people are, or are not, prepared to do. We face a crisis; we have an equation which we know to be correct; there is no real choice but to comply.

In trying to work out the implications of the carbon equation, there are two sets of facts which need to be considered: one set relates to climate; the other to fossil fuel. It is when the two sets of facts come together that the challenge becomes clear, enabling the development of workable pragmatic solutions. Accepting the need to comply with the carbon equation steers the debate away from convenient political posturing influenced by the lobbying of those with a vested interest in the fossil fuel economy. I am following the rigorous approach used by engineers.

When human safety is at risk, failure to do the maths is inexcusable. The design engineer has to be sure that the numbers in the structural analysis equations add up, for the design to be approved. No one is going to ride a rollercoaster based on a dodgy design. The same rigour should apply to how we deal with something as critical as the future climate.

Focusing on the carbon equation helps to blow away the confused thinking evident in the politics of climate change to provide a solid frame for the economic analysis. Until now, the politics of the climate debate and the associated economic analysis have only been loosely aligned with the facts. This has allowed dialogue to proceed without being called to account. Drawing the debate back to comply with the carbon equation delivers a dose of reality which may not be welcomed by those hoping to avoid facing up to climate change. For those of us who are prepared to face up to climate change, the carbon equation provides clear direction towards the solution. I will start with focusing on climate, and to keep it simple I will focus on carbon dioxide levels in the atmosphere.

# Climate change facts

## CLIMATE FACT 1

*A 'normal' level of $CO_2$ in the atmosphere is 280 ppm.*

Precise measurements of atmospheric carbon dioxide have been recorded at the Mauna Loa Observatory, Hawaii, on a regular basis since 1958. The first reading measured the atmospheric $CO_2$ concentration at 313 ppm.[26] We can go back a lot further in

---

26    29 March, 1958.

time by analysing bubbles preserved in the ancient ice beneath Antarctica. By drilling deep and cutting out an ice core, the timeline of atmospheric $CO_2$ can be plotted going back 400,000 years. During ice ages $CO_2$ levels were as low as 200 ppm; during warmer interglacial periods this increased to around 280 ppm. It is interesting to note that 80 ppm below the normal level correlates with an ice age. So, a difference of 80 ppm below normal can be significant. What if the difference goes the other way? The current $CO_2$ level is 400 ppm; a massive 120 ppm above normal. We don't have any records to show what the climate will do, because the world has not had levels of $CO_2$ this high for at least the last 400,000 years, perhaps not for the last million years.

Up until the first measurements at the Mauna Loa Observatory, industrialization had increased atmospheric $CO_2$ by around 30 ppm. 30 ppm above normal does not seem too bad, and well short of the 80 ppm which we know to be significant. Since 2000, the use of fossil fuel has accelerated, driving $CO_2$ levels higher at a rate of 20 ppm per decade to the current level of over 400 ppm. This is a whopping 120 ppm above normal. No wonder scientists are worried; we should be worried too. This is massively outside the normal range. It hardly needs the expertise of a climate scientist to ring alarm bells. It is obvious that this is not a good place to be.

The whole of recorded human history over tens of thousands of years has played out in a relatively stable and consistent climate. If a normal level of $CO_2$ could be retained at 280 ppm, there would be no climate crisis. We could be confident that the climate would remain within normal parameters. This is not where we are. Our stupidity in undermining our climate security is astounding. Each tank of fuel, each bag of cement, each blast from our central heating hardly seems to matter; but it all adds up, and there will be consequences.

Climate Fact 1, that the normal level of $CO_2$ in the atmosphere is 280 ppm, provides the baseline; showing the extent to which the fossil fuel economy has led us into dangerous territory.

## CLIMATE FACT 2

*Planet Earth has warmed by approximately 1 degree since 1800.*

According to the IPCC: 'Human-induced warming reached approximately 1°c (likely between 0.8°c and 1.2°c) above pre-industrial levels in 2017'.[27] I wanted to include the exact words of the IPCC to show how meticulous they are in being as accurate as possible. This temperature rise is a direct consequence of the fossil fuel economy. This relies on energy from outdated processes which release fossil carbon which had been locked up in the Earth's crust. We have had better technology for some time but vested interests and simple inertia have kept us hooked on energy from dirty sources. The dangers of burning fossil fuel were not appreciated when the quantities were small relative to the size of the planet. As the architects of the Industrial Revolution developed machinery such as the steam engine (Thomas Newcomen, 1712 and James Watt, 1769) and processes such as steel production (Henry Bessemer, 1856), all taking place on the tiny islands of Britain, they can hardly have understood the potential impact on the global climate. Localized pollution from burning coal was evident but it would be more than a century before the global significance was understood. By then, an economy based on fossil fuel had become firmly entrenched. As

---

27  IPCC (2019), 'Global Warming of 1.5°C. An IPCC Special Report on the impacts of global warming of 1.5°C above pre-industrial levels and related global greenhouse gas emission pathways, in the context of strengthening the global response to the threat of climate change, sustainable development, and efforts to eradicate poverty': p.31. [Available online: www.ipcc.ch/sr15/; accessed 26 July 2020].

industrialization spread throughout the world, the release of fossil carbon increased and then accelerated dramatically from the 1990s. This acceleration was due to the turbocharging of the fossil fuel economy by economic globalization. The fossil fuel industrial economy was being replicated across the planet as all countries sought to reap the rewards of economic expansion. This global phenomenon of economic expansion – of an economic model based on fossil fuel – correlates precisely with accelerating damage to the climate.

It is obvious that as we deal with economic issues this can have environmental consequences. It is less obvious, but equally true, that this can work in reverse. Dealing with environmental issues can have economic consequences. To deal with the environmental issue of climate change we need to adjust the economic model. The fact that some narrow-minded people are willing to put the economy above all else should not be allowed to drive decision-making. We should think long and hard about the importance of the environment in relation to the economy. We should demand a stable economy which does not impact negatively on the environment. I find it surprising that some people regard this as a controversial way to work economic policy. In the twenty-first century we need to up our economic policy game to move beyond short-termism to protect our long-term future.

Climate Fact 2, that the Earth has warmed by approximately 1 degree since 1800, is due to an economic model based on fossil fuel. This economy has been further turbocharged by economic globalization. Little action has been taken to change fundamental economic policy, so further increases in $CO_2$ and consequent further temperature increases are inevitable. One degree might not seem like much, but if global warming reaches 2 degrees this gets dangerous. The current path could take us to 4 degrees and even

6 degrees above normal, leading to catastrophic scenarios.[28,29] We do not need to go there; we can change direction by closing the fossil fuel economy and building a new economic model.

## CLIMATE FACT 3

*The level of $CO_2$ in the atmosphere already exceeds 400 ppm.*

The level of $CO_2$ in the atmosphere exceeds 400 ppm, which is 40% higher than normal. It is not surprising that as a consequence the climate is changing. This is not a prediction for some time in the future; this is happening now, impacting on people and communities. No single weather event can be categorically assigned as a consequence of climate change, but the trend is clear. Extreme weather is becoming more common. Storms which compared with historic records could be expected once in fifty or a hundred years, are arriving more frequently. Ancient ice in glaciers and at the poles is melting; the mean global sea level is rising. The historic records are no longer useful as we push the climate way outside what might be considered 'normal'. If we want a normal climate, we will have to return $CO_2$ to normal levels.

If our planet had a control room, such a large deviation from normal settings would activate red flashing lights and emergency alarms. The climate is changing, and could accelerate dramatically if we reach a tipping point. There is no way of knowing. We would be flipping through the Earth System Management manual for action on emergencies – except there is no such manual. We do not have a procedure to follow; we need to search out the best advice available and act on it without delay.

Climate Fact 3, that the level of $CO_2$ in the atmosphere already exceeds 400 ppm, means we are facing an emergency.

28  Lynas, M. (2007), *Six Degrees: Our Future On A Hotter Planet*, London: Fourth Estate.
29  Wallace-Wells, D. (2019), *The Uninhabitable Earth: A Story of the Future*, London: Penguin.

# CLIMATE FACT 4

*There is a time delay between levels of $CO_2$ in the atmosphere and the consequences.*

The Earth's system takes time to adjust. The climate changes we are experiencing now are a foretaste of the changes which will unfold as a consequence of current $CO_2$ levels. Even if we stopped releasing fossil carbon immediately, the climate would continue to change for many decades into the future. The extent of the climate adjustment is not known. All we know for sure is that it will be greater than we are observing now – and could be much greater. It is important to understand that although we have caused climate change, the changes which follow are outside our control. As we push levels of $CO_2$ higher it can be compared with slowly pushing a heavy railway wagon along a track with a flat gradient through heavy fog. All seems to be under control until a point is reached where the gradient turns downhill. The wagon will career off down the track into a future we cannot see or predict and end up somewhere completely different. We will be able to run after it but we will have no say in the direction it takes or where it will come to a halt. If we understood the situation in such terms, it would not take great intellect to decide that the best course of action would be to stop shoving. Even better would be to shift our effort to the other side of the wagon, first to stop it moving and then start to move it back along a direction we know to be safe. Climate change is a major calamity shaping up in slow motion, completely obvious, and no one is taking a leadership role to prevent it. If the situation was not so serious it would be like a scene in a Keystone Cops film.[30]

---

30  The Keystone Cops were fictional, humorously incompetent policemen, featured in silent film slapstick comedies produced by Mack Sennett for his Keystone Film Company between 1912 and 1917.

Climate change will continue for many decades into the future, until a new equilibrium is established over which we will have no control. If that does not cause you concern, I suggest that you have a complete lack of imagination, and are living in denial of reality. If this insight makes you break into a cold sweat and keeps you awake at night, you are not mad but entirely sane. Such a reaction would indicate that you are in touch with the reality of living through the crazy period of the early twenty-first century. Clear and honest judgment may be in short supply; but reality will have to be faced eventually.

Climate Fact 4, that there is a time delay between levels of $CO_2$ in the atmosphere and the consequences, means that the situation is far worse than we are observing now, so we need to act quickly.

## CLIMATE FACT 5

*A 'safe' level of $CO_2$ in the atmosphere is 430 ppm to have a good chance of keeping global warming under 1.5 degrees.*

The first four climate facts are precise, accurate and beyond dispute. The fifth is a matter of judgement. Those responsible for making the judgement are the world's best climate scientists. We could not have a more expert opinion. The fossil carbon emitted so far through two centuries of industrialization has brought us to a level of 400 ppm atmospheric $CO_2$. The IPCC scenarios which would have a good chance of keeping global warming below 1.5 degrees are based on not exceeding 430 ppm.[31] To hit the target agreed by the UNFCCC (1.5 degrees) means backtracking on fossil fuel dependency and stopping burning fossil fuel completely before reaching 430 ppm.

---

31 'Mitigation scenarios in which warming is more likely than not to be less than 1.5°C relative to pre-industrial levels by 2100 are characterized by concentration levels by 2100 of below 430 ppm $CO^2$...'. (IPCC 2014: 81).

Climate Fact 5, that a safe level of $CO_2$ in the atmosphere is 430 ppm to have a good chance of keeping global warming under 1.5 degrees, should focus our minds on bold and rapid action.

## CLIMATE FACT 6

*The upper limit of $CO_2$ to avoid dangerous climate change is 450 ppm to have a good chance of keeping global warming under 2 degrees.*

Two degrees has generated a lot of publicity as the limit which must not be exceeded. Two degrees is not a safe level; it will lead to significant climate change and significant impacts. The reason 2 degrees has gained momentum is simply acceptance that the world is incapable of taking sufficient action, quickly enough, to meet the 1.5-degree target. Two degrees is considered to be a feasible limit; provided climate change is treated as an emergency, leading to a robust and quick response. Depressingly, the action which is currently being discussed is neither robust nor quick. So, 2 degrees may also be dismissed as no longer feasible within the next few years. More optimistically, remaining within 2 degrees can be achieved with current technology – all that is lacking is the political will and a sense of urgency to mobilize the action required.

The upper limit of 450 ppm $CO_2$ in the atmosphere should be regarded as the backstop to any plan to solve the climate crisis. The fear which drives me to write this book, and which should be of great concern to all of us, is the alternative. The alternative is to allow climate change to enter dangerous territory of 2 degrees or more. Continued discussion and continued procrastination until 2 degrees becomes impossible leads to 'dangerous climate change' being accepted as inevitable. Another target will then be chosen, say 2.5 degrees, and described using words such as 'severely dangerous climate change'. When that target also becomes impossible a further target is adopted described as 'exceedingly dangerous climate change' followed by 'astoundingly dangerous climate change'. As the years go by and the language of extreme danger is exhausted, we could become resistant to yet further calls for action and accept the inevitable consequences of our failure to act.

Climate Fact 6, that the upper limit of $CO_2$ to avoid dangerous climate change is 450 ppm to have a good chance of keeping global warming under 2 degrees, should be accepted as fact; and not rewritten to cover political failure.

## SUMMARY OF CLIMATE CHANGE FACTS

The facts about climate change which we need to feed into the carbon equation are:

1. A 'normal' level of $CO_2$ in the atmosphere is 280 ppm.

2. Planet Earth has warmed by 1 degree since 1800.

3. The level of $CO_2$ in the atmosphere already exceeds 400 ppm.

4. There is a time delay between levels of $CO_2$ in the atmosphere and the consequences.

5. A 'safe' level of $CO_2$ in the atmosphere is 430 ppm to have a good chance of keeping global warming under 1.5 degrees.

6. The upper limit of $CO_2$ to avoid dangerous climate change is 450 ppm to have a good chance of keeping global warming under 2 degrees.

These six facts are all we need to know to decide our immediate course of action. A normal level of $CO_2$ is 280 ppm; at this level climate change would not be happening. Two centuries of industrialization had pushed $CO_2$ levels to a moderate 313 ppm by the 1960s, but the acceleration in emissions since then has taken us to an eye-watering 400 ppm. Global average temperature has increased by 1°c, starting a process of climate change over which we have absolutely no control. Emissions continue to rise; temperature continues to rise; risk continues to rise; sea levels continue to rise; our concern continues to rise; but we can only manage a limp response. We face a serious long-term challenge; we need immediate action to be able to sow the seeds of a solution. Our target is to hold $CO_2$ at 430 ppm; the upper limit to avoid dangerous climate change is 450 ppm. The action required is to close down fossil fuel extraction quickly enough to comply. Therefore, I will extend the maths to fossil fuel.

# Fossil fuel facts

The relevant facts about fossil fuel are relatively easy to define. We know how much oil is pumped and how much coal is dug, and have good estimates for the total reserves of fossil fuel already discovered. Existing oil fields which are in production are assessed using seismic measurements and flow rates to determine reasonably accurate figures for the exploitable reserves. The key relevant facts about fossil fuel are:

## FOSSIL FUEL FACT 1

*The world is emitting 37 gigatonnes (Gt) of $CO_2$ annually through burning fossil fuel.*

The world emits so much $CO_2$ because 85% of world energy comes from fossil fuel (see Table 6.1).[32]

---

32   Le Quéré et al. (2018), Global Carbon Budget 2018, *Earth System Science Data*, 10, 2141–2194.

**Table 6.1** World Energy Consumption by Source 2018[33]

| Source | Million tonnes oil equivalent | % |
|---|---|---|
| Oil | 4,662 | 34 |
| Coal | 3,772 | 27 |
| Natural gas | 3,310 | 24 |
| Hydro | 949 | 7 |
| Nuclear | 611 | 4 |
| Renewables | 561 | 4 |
| **TOTAL (world):** | **13,865** | **100** |

The world requires energy of course, but there is no requirement that it has to come from fossil fuel. Fossil fuel was the choice we made two centuries ago and it has been easy to stick with what we know. We find ourselves in a climate crisis because we have been lazy. Reducing the level of $CO_2$ emissions is possible with a dual approach. First, greater efficiencies to reduce the quantity of energy demanded. Second, ramping up other energy sources. It is perfectly possible to drastically reduce annual $CO_2$ emissions if governments orchestrate change to the energy economy. There is no need to continue emitting $CO_2$ at this level, but little will change until we alter the economy.

Fossil Fuel Fact 1, that the world is emitting 37 Gt of $CO_2$ annually through burning fossil fuel, is a staggering amount and only continues because we remain committed to an economy based on fossil fuel.

## FOSSIL FUEL FACT 2

*At current rates of fossil fuel consumption, the world will have emitted the total allowable quota of $CO_2$ to hold to the global*

---

33  BP (2019), BP Statistical Review of World Energy 2019, 68th edition [availble online: bp.com/energyoutlook: p. 9; accessed 26 July 2020].

*warming target of 1.5 degrees by the year 2034.*

This fact is deduced by taking the difference between the current $CO_2$ level in the atmosphere (400 ppm) and extrapolating current emissions until the 1.5-degree limit is reached (430 ppm). At current rates of consumption, we will reach 430 ppm in less than 15 years, around the year 2034. At this point, the burning of fossil fuel will have to cease if we are to limit global warming to 1.5 degrees. In reality, phasing out fossil fuel cannot be instantaneous; it will take time and needs to start without any further delay. With bold action to slash emissions quickly in the next few years, the date when fossil fuel consumption would have to end could be pushed back beyond 2034. The faster we act, the easier it will be.

Fossil Fuel Fact 2, that at current rates of fossil fuel consumption, the world will have emitted the total allowable quota of $CO_2$ to hold to the global warming target of 1.5 degrees by the year 2034, shows the urgency with which we must take action.

## FOSSIL FUEL FACT 3

*There are huge potential reserves of fossil fuel.*

Coal is available in abundance and could last for centuries but coal is not an attractive fossil fuel because of its high carbon emissions. We have enough of the lower-carbon fuels, oil and gas, for our current needs. Some investment has flowed into exploiting unconventional sources such as shale rock containing gas and tar sands such as those in Alberta, Canada. The energy used to extract them means their carbon emissions are greater, so again these are not attractive fossil fuels. If oil and gas prices shoot high, and look likely to stay high, then those who continue to believe in the fossil fuel economy will be tempted to ramp up investment in these unconventional sources but it will be high risk. We are already seeing investors pulling out of coal and banks are reluctant to lend to coal projects. This reluctance towards new investment will extend to unconventional fossil fuels once it becomes clear that governments will finally face up to climate change and take action. Overall, there are plenty of reserves of fossil fuel so no prospect that fossil fuel will run out. The shortage will be of investment capital to build the facilities to extract it as the world faces up to climate change.

The fossil fuel industry will close down through a lack of investment capital as governments discourage its use rather than because reserves run out. As John Elkington so neatly expressed it, the Stone Age didn't end because they ran out of stone.[34]

Fossil Fuel Fact 3, that there are huge potential reserves of fossil fuel, means that the availability of fossil fuel is not a prime factor in closing down the fossil fuel economy.

## FOSSIL FUEL FACT 4

*There is a hierarchy of fossil fuel from the cleanest, which is gas, followed by oil, and the dirtiest fuels which are coal and unconventional oil.*

**Table 6.2** Fossil Fuel $CO_2$ Emissions Factors[35]

| Fossil fuel | $CO_2$ emissions factor (Kg $CO_2$ per MWh of energy) | Comparison with gas |
|---|---|---|
| Natural gas | 202 | 100% |
| Crude oil | 264 | 76% |
| Coal | 360 | 56% |
| Oil shale and tar sands | 403 | 50% |

---

34 Elkington, J. (1997), *Cannibals with Forks: The Triple Bottom Line of 21st Century Business.* Oxford: Capstone.

35 Source of data in Table 6.3: H. Ritchie and M. Roser (2019), 'Fossil Fuels', OurWorldInData.org [Available online: https://ourworldindata.org/fossil-fuels; accessed 26 July 2020].

Fossil fuel should not be considered a single category. The carbon calculation makes it clear that there is only so much fossil carbon which can be released into the atmosphere before dangerous climate change is initiated. The simple conclusion is that the extraction of fossil fuel should cease. This is what will have to happen, but any pragmatic analysis would accept this cannot be done overnight. Assuming fossil fuel continues to be used for a period of time it would be inefficient to use fuels with the least energy return for the carbon emitted. Approximately twice as much energy can be obtained using gas compared with coal. For the coal industry, the death knell has already sounded. Investors are steering clear and only populist politicians seeking to appeal to voters in coal mining areas are responding positively to lobbying from a dying industry. It is not surprising that the industry tries to stay solvent but industry insiders know as well as anyone that the window to continue making profits is closing fast.

It is worth noting that unconventional oil is one of the least efficient fossil fuels. Through the lens of the carbon equation, knowing that there is a limit to the total fossil carbon we can emit, the logic says that the unconventional oil industry should be regulated out of existence in its infancy. The death of coal is widely understood and accepted, but the need to stop the exploitation of tar sands and oil shales is not widely appreciated. Opposition to these industries tends to come from environmentalists concerned at the direct environmental impact. There should be no need for such protest; the logic against unconventional oil is solid on the basis of carbon emissions alone.

Fossil Fuel Fact 4, that there is a hierarchy of fossil fuel from the cleanest which is gas followed by oil, and the dirtiest fuels which are coal and unconventional oil, means that the remaining allowable carbon budget should be used for the cleaner fuels.

# FOSSIL FUEL FACT 5

*The total known reserves of conventional oil and natural gas would use up the allowable carbon budget.*

**Table 6.3** Oil and Gas Reserves
Compared with Allowable Carbon Budget[36]

| Global warming limit | Total $CO_2$ budget (Gt) | Proportion world gas reserves consumed | Proportion world oil reserves consumed | Remaining $CO_2$ budget (Gt) |
|---|---|---|---|---|
| 1.5 degrees | 580 | 100% | 40% | 0 |
| 2 degrees | 1,000 | 100% | 100% | 275 |

The recommended target for halting global warming at 1.5 degrees would be reached by consuming all known reserves of natural gas and about 40% of known oil reserves. Turning to the upper limit to avoid dangerous climate change, burning all known reserves of natural gas and oil takes us close to the limit. Bearing in mind that the world will need a number of years to get its act together, the remaining $CO_2$ budget is likely to get used by countries which struggle to implement a rapid closure of coal.

Burning the total known reserves of gas and oil would smash the carbon budget for 1.5 degrees and require most of the budget to remain within 2 degrees. The logical conclusion is that extraction of high-carbon fossil fuels should cease so the budget can be used efficiently. It would also be pointless to carry out exploration for new oil and gas fields. The fact that oil and gas companies continue to spend on exploration indicates that the political

---

36 To calculate these figures, the following factors have been used: World natural gas reserves ($Gm^3$) 187,000; average figure for burning natural gas 2 kg $CO_2$ per $m^3$. World oil reserves (Gbarrels) 1,700; average figure for burning crude oil 317 Kg $CO_2$ per barrel.

process has failed to put in place sufficiently robust policy to turn off investment in expanding fossil fuel extraction. The big fossil fuel companies are not stupid and will only invest when they are confident of generating a profit. Big oil corporations are targeted by environmentalists but they should not be blamed. The fault lies with governments failing to regulate, and all of us for failing to demand that they do so.

Fossil Fuel Fact 5, that the total known reserves of conventional oil and natural gas would use up the allowable carbon budget, means that all other sources of fossil fuel should be closed down forthwith.

## SUMMARY OF FOSSIL FUEL FACTS

1. The world is emitting 37 Gt of $CO_2$ annually through burning fossil fuel.

2. At current rates of fossil fuel consumption, the world will have emitted the total allowable quota of $CO_2$ to hold to the global warming target of 1.5 degrees by the year 2034.

3. There are huge potential reserves of fossil fuel.

4. There is a hierarchy of fossil fuel from the cleanest, which is gas followed by oil, and the dirtiest fuels which are coal and unconventional oil.

5. The total known reserves of oil and natural gas would use up the allowable carbon budget.

The world is pumping $CO_2$ into the atmosphere at such a high rate that we need to stop burning it within 14 years if we are to stay safe. There are huge reserves of fossil fuel so rather than wait until supplies run down, we will have to close down extraction. The transition from the current economy to the future economy will take time. During the transition, we should cut energy demand and ramp up clean energy sources; and where we continue to burn fossil fuel it should only be gas and conventional oil. All other sources of fossil fuel should be closed down without exception.

# Implications of the carbon equation

There are two version of the carbon equation. The first defines the target recommended by the UNFCCC for limiting global warming to 1.5 degrees:

$$580 \text{ billion tons } CO_2 = 1.5 \text{ degrees}$$

The second defines the upper limit of total $CO_2$ emissions to avoid dangerous climate change:

$$1,000 \text{ billion tons } CO_2 = 2.0 \text{ degrees}$$

It is widely feared that the 1.5-degree target is not now achievable, so taking the pragmatic approach, I will work out the numbers to keep within 2 degrees. The following discussion is therefore about action which *must* be taken to keep within limits which *cannot* be breached. The world faces a crisis; the equation is clear; failing to comply is not an option.

What does the equation tell us about climate change and fossil fuel? What deductions can we make? If examination of the equations provides uncomfortable truths and exposes inconvenient implications, these need to be accepted and made use of rather than their veracity attacked. Quantifiable facts (which we know to be accurate) and the equations which connect them (which we know to be logically correct) are the basis of sound analysis. For engineers, this is the only way to build structures which will stand up and survive. For policymakers, this is the only way to craft policy which stands up to examination and has a chance of succeeding.

Applying the carbon equation exposes difficult truths. This is the nature of the situation and cannot be wished away. The dangers of climate change and society's addiction to fossil fuel is a lethal combination. Such lethality requires a robust and rapid response which is in accord with the facts using maths which adds up.

The carbon equation tells us the sum total of $CO_2$ emissions to cap global warming to 2 degrees. This equates to burning known reserves of conventional oil and gas. Gas is the cleanest fossil fuel providing the most energy for its associated $CO_2$ emissions; oil is next, with coal and unconventional oil the dirtiest and least efficient fossil fuels. The best way to use the allowable carbon emissions 'budget' would therefore be to burn only gas and oil. To keep within the budget, only the known reserves would be exploited until they were depleted. This means accepting that the current gas and oil industry is allowed to continue to operate current fields. However, there is no space in the budget to open new fields. The world would have to accept that supplies of fossil fuels would become increasingly tight as conventional oil and gas supplies run down. It is important to avoid a knee-jerk reaction to deny what the equation tells us. Thinking that it would be impossible would be the response of the fossil fuel addict. The better response is to see with great clarity the need to invest in renewable energy, energy efficiency and support the development of an economy beyond fossil fuel. It actually makes perfect sense to cap fossil fuel extraction knowing that this will facilitate the transition required; but making sense does not make it easy. Focusing on how to establish the political and economic framework to limit fossil fuel extraction takes the analysis into examination of what is possible.

I have the freedom in writing the analysis presented here to go wherever the logic leads. There comes a point when theory makes contact with pragmatic reality. Reflecting on the feasibility, in political and economic terms, of closing down existing oil and gas facilities, does not look to be possible. The investment has already been sunk in exploration and building the infrastructure to exploit existing fields. There will be on-going maintenance costs, but such oil and gas will flow for many years at low cost and require little substantial new investment. Oil and gas are also the cleanest fossil fuels. Pragmatic economic and political reality, matched with the hard logic of the carbon equation, means accepting that current conventional gas and oil fields will continue to operate until depletion – or until fossil fuel enters the end game in the years ahead. So far, so good. This implication of the carbon equation is palatable and easy to digest; and easy to sell. There is no need to close down existing conventional gas and oil fields already in production.

The politics of trying to close down fossil fuel extraction will be tricky. The pragmatic approach outlined here is that with regard to existing oil and gas facilities, we do not try to close them down. This will conserve valuable political capital to expend on closing down the dirty fossil fuels. Environmentalists should learn a lesson here. Opposing the big oil and gas companies is a waste of effort. Opposition should focus on the dirty fuels, to speed up the elimination of coal and to kill off oil-shale and tar-sand exploitation.

The recommendation that existing conventional oil and gas fields should continue without interruption does not sit in isolation. This only fits the carbon calculation by completing the logical analysis. The flipside, and most difficult implication, is that there is no room within the allowable carbon budget to burn any more coal or to exploit unconventional sources such as oil shale and tar sands. The benign approach to conventional oil and gas has to be linked to a robust approach to closing down coal and unconventional fossil fuels.

# Country perspectives

Each country will have its own challenges in complying with the carbon calculation in ways which fit their circumstances. The fossil fuel economy has become the default option, even in places where it makes no sense. There are countries which have ample solar renewal energy capacity, and little or no fossil fuel reserves, which have been advised by outside 'experts' to develop in the image of Western industrialization, perpetuating the idea that the fossil fuel economy is the best model to copy. This outdated view has to be replaced by deep expertise in building low-carbon economies. Escaping the grip of fossil fuel addiction could hit some countries' economies hard. In many cases, unilateral action would be an economic own-goal. All countries who accept the carbon calculation should therefore work together in negotiating an international fossil fuel agreement (this is the subject of the next chapter). Below I will explore a range of other national perspectives to get a flavour of the breadth and variety of the political and economic challenges ahead.

Some countries, such as the UK, will find it relatively easy to comply with the carbon equation (see box).

The UK might be willing to take a leadership role on the world stage because the country is already moving towards closing down the extraction of the dirtiest fossil fuels.

For other countries, accepting the implications of closing down the dirty fossil fuels may be a huge challenge requiring considerable investment and realignment of the economy. Legislation can be used to change the commercial reality and make the transition, but all too often politics gets in the way. In the United States, there is a lively debate between populist politicians mounting a last-ditch defence of the coal industry, and state legislators acting against dirty fuels as cleaner energy

**The United Kingdom**

The UK has coal, oil and gas, but the transition away from fossil fuel has begun. The coal industry was the country's main source of energy until the 1960s but since then there has been a steep decline. The last remnants are limping along but the industry is almost dead. There will be little political resistance to finally closing it down completely. Oil and gas from under the North Sea continues to flow but fields are starting to reach the end of their life. The UK also has shale rock which can be fracked to extract gas. As supplies from the North Sea tail off, there is some support for fracking to maintain supplies. The fracking industry has been keen to invest but is becoming spooked by widespread public concern over the environmental impact of extraction, such as earthquakes and risks to water supplies. This is where well-targeted public protest has been influential. As anti-fossil sentiment grows and as the government tightens regulations, the UK fracking industry is being killed off before it gets going. In general, the UK would find it relatively easy to comply with the carbon equation. The three factors which are coming together are: public concern, government regulation, and investors cutting their losses.

sources become increasingly affordable. The United States has the potential to be a leader in developing a low-carbon economy but at the time of writing, with Donald Trump in the White House, the United States is being left behind.

Australia is an example of a country which could do well from shifting to a low-carbon economy, but seems to be trapped by its politics. The coal lobby has a strong influence and coal is a valuable export. To face up to climate change and turn away from coal will require courage and conviction. The truth is that it would be relatively simply to close down the Australian coal industry as the country has almost unlimited solar energy capacity (see box).

## Australia

Australia has huge deserts with considerable capacity for harvesting solar energy. This could be used directly to replace fossil fuel in power generation. There is also the potential to build facilities to manufacture 'liquid sunshine'.[37] I coined this term to avoid specifying its exact nature. I mean a hydrocarbon fluid, which can be used as fuel, that is easily transported, and produced using facilities powered by the Sun. An example is transparent water tanks containing algae which use sunlight as part of the process which makes biodiesel. Exports of liquid sunshine could be much more valuable, and sustained over the long term, than the income from exporting coal. Australian politicians would have to face down the strong coal lobby in order to face up to climate change, but the long-term outcome of doing so looks positive.

Australia could be a world leader in renewable energy; but Australian politics seems to be stuck in the past with deeply ingrained reluctance to abandon the fossil fuel economy. Whether Australia would decide to ban coal unilaterally is doubtful, unless this move was part of a wider global agreement. This is why global agreement is so important.

In Europe, Germany is making progress to transition away from coal but countries such as Poland remain highly reliant. Perhaps the EU could adopt a policy, backed up by support mechanisms, to close the coal industry across Europe. Whether the EU would do this unilaterally is doubtful, unless as part of a wider global agreement. This is why global agreement is so important.

---

37  McManners, P.J. (2008), *Adapt and Thrive: The Sustainable Revolution*, p. 68, UK: Susta Press.

Canada is an important test case for fossil fuel policy. How Canada decides to respond could be crucial to setting the tone for a worldwide approach (see box).

---

### Canada

It is estimated that Canada's tar sands may contain as much oil as Saudi Arabia. Canada could become a major world oil producer, but the bituminous oil contained within the sand is hard to extract. Considerable processing, heat and energy is needed to separate the oil from the sand. When all this is added up, the overall carbon emissions are on a par with coal. This is quite different to Saudi Arabian crude oil which flows freely almost like opening a tap.

Canada has, in the oil sands of Alberta, the potential to become the emissions monster of fossil fuel. Building the Keystone Pipeline System between Canada and the United States (commissioned 2010) shows Canada's current intent to exploit the tar sands. If this changes, the pipeline's owner, TC Energy, would be mightily disappointed to find that their asset becomes worthless. This would be the correct course of action, which it can be argued TC Energy should have anticipated.

---

There is a debate taking place in Canadian politics between those who would like the country to become a major oil producer and those who understand the global environmental consequences. Whether Canada would decide to ban oil extraction from tar sands unilaterally is doubtful, unless as part of a wider global agreement. That is why global agreement is so important.

For fossil fuel addicts, short-sighted economists, and selfish politicians only interested in the electoral cycle, the implications of the carbon equation will be hard to accept. However, the equation speaks truth to those in power. Although it may be politically difficult to close down coal and kill off development in unconventional oil and gas, the rationale for doing so is sound. If this logic is matched by public opinion demanding action, the politics could switch, making closing coal a political necessity. On a positive note, deciding to implement the logic which flows from the carbon equation will mobilize investment in building the infrastructure for a new and different economy. If fossil fuel addicts can kick the habit, economists give greater weight to the long-term, and politicians adopt a more principled approach, the changes required are feasible.

To explore how the carbon equation might apply to a developing country, let us consider India.

### India

India is one of the world's most populous nations with considerable potential to grow its economy. How India does this is important. If it does so by ramping up a fossil fuel economy there will be a massive impact on the global climate. India is currently highly reliant on coal so it will have to be persuaded to leave the coal in the ground. However, India is at a relatively early stage of development and does not have extensive fossil fuel infrastructure. India could move on past the dirty stage of industrial development and leap forward beyond the old model of an electricity grid fed by coal-fired power stations. It would be in the world's interest to allow India access to appropriate technology at low cost to drive such a leapfrog transformation. The solution could be local power supply solutions based on renewable energy, particularly solar but also wind.

Developing countries like India face many challenges to improve the lives of their people. It will require an appropriate global policy framework to ensure that it is in the interests of all countries, including developing countries, to join an international push to close down coal.

## Conclusion

The carbon equation is right at the heart of facing up to climate change. Accepting its logic and identifying its implications exposes the reality of the situation. The allowable carbon budget will be used up by burning all the existing reserves of oil and natural gas. This supports allowing oil and natural gas to be consumed as the economy transitions from a fossil fuel economy to a new sustainable economy. This easy decision has to be matched with a tough approach to closing down the coal industry and preventing oil-shale and tar-sand exploitation.

The analysis presented here is the way to break through the debate over climate to make realistic progress. The hard reality the equation exposes may not be welcome in some quarters, but such examination starts to build confidence that the climate challenge is solvable. Where focusing on carbon has proved to be ineffective – allowing all forms of fossil fuel to continue to be exploited – a sharp focus on fossil fuel shows the way forward. Exactly how to do this is the subject of the next chapter.

*Climate protest gets real*

# CHAPTER 7

# CONVENTION ON FOSSIL FUEL (COFF)

It is now urgent that the world orchestrates joint action to deal with climate change. Establishing a convention on fossil fuel is the only realistic way forward. Such a mechanism would expose the dilemmas which governments need to navigate in order to face up to climate change. Where the UNFCCC has been sidetracked by a range of climate-related issues, the proposed Convention on Fossil Fuel (COFF) would focus on the cause of the problem. It is important to go beyond general demands to do something about climate change, to campaign for specific and effective action to close down the fossil fuel economy. Initiating such a convention would bring clarity and open the way to effective coordinated action.

How the convention on fossil fuel is set up, and the task it is given, will dictate how it operates and how likely it is to succeed. I propose that a suitable remit for the COFF would be:

*The convention on fossil fuel is charged with considering how fossil fuel extraction can be closed down in line with a maximum total level of $CO_2$ in the atmosphere of no more than 450 ppm.*

The exact structure of the convention and how it carries out its work will need to be decided. An example from which we can draw inspiration is the Vienna Convention for the Protection of the Ozone Layer. This is an example of long-standing concern by scientists reaching a pivotal stage where fast action followed. Ozone has a vital role in protecting us from harmful ultraviolet radiation. Through the 1970s scientists had become concerned that fluorocarbons were damaging the ozone layer. The incident which made the world wake up occurred in 1985 when scientists from the British Antarctic survey reported that they had detected a hole in the ozone layer over Antarctica. Its discovery was announced in a paper in the journal *Nature* by British Antarctic Survey's Joe Farman, Brian Gardiner and Jonathan Shanklin.[38] The danger was further exposed by examples such as sheep in New Zealand being blinded by ultraviolet radiation.[39] These warnings had urgency and were seen as a clear and present danger to humanity. The world came together and responded with great speed. The Vienna Convention for the Protection of the Ozone Layer was agreed, establishing a framework for negotiating international regulations on ozone-depleting substances. On 15 September 1987 the Montreal Protocol on Substances that Deplete the Ozone Layer was adopted. At the time of writing, the Protocol is the only UN treaty that has been ratified by every country on Earth – all 197 UN member states.

The Vienna Convention was driven by concern over the ozone layer, but it did not focus on the ozone layer; it focused on regulations to control ozone-depleting substances. There is an important lesson here for dealing with our concern over the climate. The convention should not focus on climate but it should focus on controlling climate-damaging substances. The ozone hole was accepted to be a crisis leading rapidly to an agreement to act. Accepting climate change as a crisis could go the same

---

38  Farman et al, (1985), Large Losses of Total Ozone in Antarctica Reveal Seasonal ClOx/NOx Interaction, *Nature*, 315, 207–210.

39  Goldfrank, W.L, A. Szasz and D. Goodman (eds) (1999), *Ecology and the World-system (Contributions in Economics & Economic History)*, p 28, USA: Greenwood Press.

way. The Vienna Convention delivered action within two years; the COFF could deliver in a similar timescale.

The situation we now face with regard to climate change has moved much closer to the circumstances in which the Vienna Convention was established. Excess carbon dioxide emissions from burning fossil fuel are now regarded as a clear and present danger to humanity. After many years of research and development there are now viable and affordable alternatives. Renewable energy technologies are mature and, as they scale up, costs have been falling. There are also massive improvements in energy efficiency such as LED lighting, which needs a tenth of the energy of the old incandescent lights that were standard in the twentieth century. For the world to agree to phase out fossil fuel is now technically feasible, economically viable and, crucially, the window to political acceptability is starting to open. Over the last decade the world has shifted from being unable to operate without fossil fuel to being able to make the choice to move beyond fossil fuel. The only question remaining is how quickly an appropriate protocol can be agreed and how rapidly it can be implemented. There are no real barriers to achieving an agreement except reluctance to get on with it.

It seems to me that we don't see the window of opportunity because we are so firmly trapped by the hope that the UNFCCC is on the case. When UNFCCC is put aside, and the situation considered afresh, the insight is both disturbing and inspiring. It is disturbing in exposing complete inaction; it is inspiring in that simply deciding to face up to the climate crisis is enough to make the solution feasible. We have the technology and deploying it is economically viable; it is only attitudes which are slow to change. Although people reliant on the fossil fuel industries will need to seek employment elsewhere, the economy beyond fossil fuel generates far more jobs than will be lost. For the first time in the

climate crisis, all the key factors are lining up. The only opposing forces are political inertia and lobbying from the fossil fuel
industry. These are powerful forces but they can be overpowered
where there is the will to do so. The fossil fuel industry has generated employment and wealth for many countries but politics is
shifting towards understanding and accepting that this cannot
last. Now that investment is leaving the fossil fuel industry (in
response to both public protest and investors safeguarding their
funds) the industry's lobbying power is starting to drain away.
This dawn of political and investment reality is an opportunity
which must be grasped.

The success of the Vienna Convention can be replicated if
it is acknowledged that we face a climate emergency and insti-
gate a process which is similarly rapid and robust. We have a
head start, in that the science is well understood and documented. The COFF can push forward to work on the options to
respond. Through open debate, the options can be considered
and the final protocol agreed. Rapid adoption by all 197 members of the UN may not be guaranteed because some countries
will find it disproportionately hard to comply. That is why the
protocol will have to consider measures to facilitate the transition, as well as controls which close down fossil fuel extraction.
I hope that the COFF can deliver the second UN treaty to be
ratified by every country on Earth. The need is great and the
circumstances are right. We are no longer threatened by damage to the ozone layer. There should be no need to be threatened
by climate change.

The COFF will have a mandate to close down fossil fuel. Its
mission is clear and the convention must be allowed the space
for objective analysis and thinking through the consequences
without being constrained by national self-interest. At its core,
the COFF needs a robust analytic process; this will be the key

to success. When proposals from COFF are presented to wider debate, then national self-interest can be expected to surface as nations defend their position. The maintenance of a strong analytic process will ensure that the pleading of vested interests is exposed for what it is. The protocol agreed can be expected to require radical change to dismantle the fossil fuel economy and replace it. It is unavoidable that there will be winners and losers. It is hoped that the world's leading nations will take a responsible and pragmatic approach. Where it is clear that a particular country will be a loser, they should not be allowed to veto the process or seek to undermine the analysis. It would however be entirely reasonable and sensible that such a country engages in discussion about how to mitigate the downside. It is in the interests of all countries to broker a solution; give and take should be integral to the process. Ultimately, there may be countries which fail to engage fully with the process and may be reluctant to sign an agreed protocol, judging that the short-term negative consequences are too great. In such circumstances, a coalition of the willing should drive the process forward leaving foot-dragging countries behind. This is to be expected and should not be allowed to slow progress. It is not unusual in international politics for coalitions of countries to push forward knowing that to be engaged in the process means to be influencing the direction. Other countries tend to come on board fearing that the outcomes might be unfavourable if they are not at the table. Maintaining momentum is the key so that remaining outside looks increasingly risky.

## Timescale

The first time that the world came together to consider climate change as a real threat was the United Nations Conference on

Environment and Development (UNCED), Rio de Janeiro, 3–14 June 1992. This is often referred to as the Earth Summit of 1992. Looking back at that point in history, it would seem that although scientists and environmentalists were concerned, world leaders generally were not. The general public were not aware of how potentially serious climate change would become, so there was no widespread demand for action. Politicians only have so much bandwidth, and it soon gets taken up by priority demands from the electorate. Politicians were therefore content to agree to the relatively simple task of convening the Earth Summit of 1992. As far as world leaders were concerned, at the time, this was all that was required of them.

The Earth Summit started a process which established the UNFCCC and the first 'success' was the Kyoto Protocol. This laid down aspirations which did not demand action or have any enforcement mechanism. It was therefore easy for world leaders to sign up. It also applied only to developed countries so developing countries such as China and India did not need to worry that their industrial expansion might be curtailed. As explained in the previous chapter, the UNFCCC has continued to chug along, meeting, discussing, writing reports, reporting 'progress' whilst achieving absolutely nothing of substance. Not only is the process flawed, but the timescale continues to extend putting off action to some indeterminate point in the future. This is quite ridiculous. The timescale of the Vienna Convention is a much better example. From recognizing the ozone hole as an emergency to agreeing an effective treaty to deal with the problem took two years. We can do the same with fossil fuel if we consider it an emergency and enact the bold measures required.

Here below I will outline what could be achieved if world leaders accepted climate change as an emergency which puts lives at risk and needs to be dealt with urgently.

# PHASE 1 – SETTING UP THE COFF

The United Nations would be the appropriate body to set up the COFF. It is the only body that has the legitimacy to act on the world's behalf. In setting up the COFF, the UN would have to be careful that it had the credibility to deal with political and economic reality. There would be little point in yet another body which produces reports which are filed and largely ignored. A small directorate of suitably qualified permanent staff will be needed to lead and administer. They would be the architects of the process; professional people who behave in a neutral manner without showing fear or favour to any nation.

Interestingly, when you reflect, the core COFF team does not need to include environmental experts. The COFF is not about climate change but about how to wean the world off fossil fuel. The COFF needs people who can engage right at the heart of politics and business to listen, and be listened to, by politicians, policymakers and business leaders.

## PHASE 2 – SETTING UP THE FOSSIL FUEL PANEL OF EXPERTS (FFPE)

A panel of experts should be set up with a range of skillsets drawn from across the world. The prime initial aims for the panel will be to define how much fossil fuel can be exploited, and which fuels to allow through the transition and which to close down without delay. The panel should include scientists, economists, business experts and representatives of NGOs. Expertise will be needed in all stages of fossil fuel from extraction to end-use, including knowledge of the direct and indirect environmental impacts. The business experts should be drawn from both the old fossil fuel economy and those experts in how to build the new economy which will replace it. Although care should be taken to avoid recruiting people whose behaviour is defined by blinkered defence of the fossil fuel industry, forward-looking oil and gas experts who understand that they are managing the final stages of a dying industry should be welcomed. NGOs with a direct interest should be invited in, not only to hear their view but also so they clearly understand what is happening. NGOs who are wary could block progress, whilst well-informed NGOs could have an important role in growing support for a treaty on fossil fuel.

## PHASE 3 – THE WORK OF THE COFF

The COFF and its panel of experts is designed to focus on the clear objective of closing down fossil fuel extraction in a timely manner appropriate to holding climate change within safe limits. Working from such an objective, the logic is clear. Provided the logic is allowed to play out, without being confused or diluted by those with a vested interest in frustrating the process, a solution is within reach. We should not expect the outcome to be the perfect environmental solution nor expect to be able to maintain the current economy. The outcome should be a convention which would dismantle fossil fuel dependency in a coordinated and robust manner.

In an ideal world, where everyone behaves for the common good, there would be no opposition to the logical analysis of the COFF. In the real world, there will be numerous parties who feel threatened and wary of what a dose of clear hard truth might deliver. These range from the fossil fuel industry, of course, and countries whose wealth derives from fossil fuel, to poor countries who feel they might be denied the chance to develop because the historic industrial development path followed by rich countries is closed down. Clarity will be needed that rather than closing down development, the changes required will steer countries along a different and better development path. In fact, countries without an extensive fossil fuel infrastructure will be better placed to navigate development beyond fossil fuel. This is because they will not have been locked into fossil fuel dependency, so will be free to set policy and channel investment unencumbered by an obsolescent industrial legacy. If the work of the COFF is carried out in a transparent manner, to an agenda which is clearly for the world's common good, it should be able to work without being held to ransom, allowing sensible and logical proposals to emerge.

# An outline of possible progress

The process established under the proposed COFF will not only identify what is to be done, but also operate in a way which has credibility and can garner widespread support. We should wait to see it established and then see what emerges. However, it is useful at this formative stage to understand the potential value the COFF could deliver. The outline which follows is merely a straw man to demonstrate what could unfold.

The logic presented is straightforward, and in some respects obvious once you accept that the focus must be on fossil fuel. This could happen within the space of just two years. Each stage will be subject to a robust analysis by experts, so these stages are only draft possible findings. My expectation is, that after setting up the COFF, it would run along the lines outlined below. I am happy to concede that the experts might find a better path and that certain aspects of the path outlined below will not turn out as predicted. The process of writing this down is to show that there is at least one path out of the climate crisis; whereas currently there is not an identified path that leads to an acceptable solution.

Whether *my* path is the one adopted is beside the point. The path proposed shows that it can be done; and giving confidence that it can be done is all that is needed at the outset to establish the COFF and set it to work.

## STAGE 1 – FOSSIL FUEL BUDGET

The first task is to decide the fossil fuel budget corresponding to the amount of $CO_2$ that the world can pump into the atmosphere. This will be the foundation for the stages which follow. I propose a tentative draft budget based on the statistics presented in Chapter 6. The experts would need to check the figures as I admit that my approximations are nothing more than a rough sarting point.

Humankind has an allowable capacity to liberate between 580 and 1,000 gigatonnes of $CO_2$ into the atmosphere. The lower figure corresponds to the target of 1.5 degrees; the upper figure to the 2-degree limit. The cleanest fuels will generate the most energy from this allowance. So, to get the most out of this last stage of the fossil fuel economy it would be sensible that the fossil fuel budget is based on the cleanest fuels, oil and particularly gas. The simple arithmetic in the previous chapter showed that consuming all the current known reserves of natural gas and 40% of oil reserves hits the lower target. Using all known gas and oil reserves gets very close to the upper limit. It is therefore likely that the initial budget would accept current reserves of gas and oil as fuels to use through the transition. The facilities for existing gas and oil fields are already in place so little additional investment would be required. This is a convenient initial conclusion which will be welcomed by the fossil fuel industry, the energy business and end-users. So, there is no need for an immediate crunch point.

The convenient conclusion to continue using existing oil and gas facilities has a flipside which is far more controversial and potentially difficult. It is that other sources of fossil fuel will have to be closed down forthwith. This instantly hits a reality check, as an immediate cessation is not going to happen. The actual budget proposed will therefore have to include some short-term use of coal and perhaps unconventional fossil fuels until they can be closed down. As with the successful Vienna Convention, this should be according to an emergency timescale of just two years, or in this case because of the greater complexity perhaps five years. An emergency requires urgent action.

There are implications for policy which should accompany the budget and lay the foundations for the next two stages. First, to focus future consumption on gas and oil will require closing off new investment in coal and unconventional fossil fuel extraction. This should be relatively easy because as soon as investors see policymakers are serious, they will not want to invest. Second, existing capacity to exploit coal and un-conventional fossil fuels will have to be closed down. This will be tough because the businesses involved are likely to be bankrupted. At the initial stage, the potential difficulty with implementation should not obscure the logic of the recommendation. These two initial implications, to starve the fossil fuel industry of investment and close down capacity in the dirtiest fuels, are clearly correct, even though there might be concern about hitting tight timescales. In Stage 1, strategic recommendations along these lines should be agreed and allowed to set the foundation for the next stage.

## STAGE 2 – CLOSE DOWN NEW INVESTMENT

Stage 2 will get into the detail and address the difficulties of implementing the fossil fuel budget. I anticipate that discussion will focus on a draft agreement to close down new investment in fossil fuel extraction.

Agreeing policy to close off new investment in the least efficient fuels, coal and oil from sand or shale, should be straightforward.

As we start to escape the straitjacket of the old economic reality and begin the transition to an economy without fossil fuel, agreeing such policy liberates investment capital to ramp up renewable energy. I suggest that there will be little need for rules to prevent new investment, because investors will get out as soon as they can without losing their shirts. Investors who already have stakes in coal and unconventional fossil fuel extraction, or banks with loans secured against such assets, should be very worried. Simply reading the analysis I present in this book, if it looks like it might be adopted, will be enough to close down investment. The investment community may not be environmentalists but they do take enormous care not to lose their money. If enough countries agree to outlaw new coal and unconventional fossil fuel, investment will dry up, even before such an agreement has been ratified by all countries.

For the cleaner fossil fuels (natural gas and conventional oil), I imagine an agreement in principle could be reached for no further exploration. The sticking point is likely to be enforcement. Winning the argument for such draconian restrictions would seem to be difficult from today's perspective whilst we are still in the grip of the fossil fuel economy. A way around this could be an agreement that any country allowing new oil and gas facilities is locked out of the world market. So, a country could make a decision to stay outside the agreement, and within its own borders, under its control, exploit new gas and oil for internal consumption. Whether they could find investors willing to finance such projects is doubtful, but a government resolutely opposed to the above-mentioned agreement might choose to make the investment (if it had the resources and expertise to do so). Other ways to force or encourage compliance could be considered in due course. For now, we should not get ahead of ourselves and ensure first and foremost that there is agreement in principle to close down exploration and investment, particularly with regard to the least efficient fossil fuels.

## STAGE 3 – CLOSING EXISTING FOSSIL FUEL EXTRACTION FACILITIES

The next stage in the work of the COFF would be to examine existing production facilities for coal and unconventional oil extracted from shale or sand. When the world regards climate change as an emergency, it would be expected that the conclusion could be reached quite quickly that these facilities should be closed down. Whether such an obvious conclusion would translate into rapid action is less certain. We should not underestimate the tenacity of countries and corporations to defend current ways of operating. What should happen is that coal and unconventional fossil fuel extraction are wound down as soon as it can be arranged. Any delay would mean wasting the allowable fossil fuel budget on inefficient fuels.

As soon as the world's determination to close existing facilities seems to be real and likely to succeed, the business case for coal-fired energy generation will collapse. Currently, a coal-fired electricity plant might have many years of operating capacity left according to the age of its machinery and its expected operating life, but almost overnight, as sentiment switches, it could face closure.

I would expect the panel of experts to examine alternative scenarios. Shorter timescales are better for the climate but politically difficult. Longer timescales delay the inevitable changes and make inefficient use of the remaining carbon emission allowance. The scenarios presented might range from almost immediate closure within one year to a slow transition over five years. I write this at a time when governments are escaping tough choices by setting distant targets such as 2050. This is simple political expediency, providing plenty of wriggle room so as not to have to do very much very soon. Currently, closing existing facilities within such a short period is seen as unachievable; not because it can't be done but because the economic consequences are significant. Once we truly adopt an emergency planning perspective, the timescales I propose here of between one to five years look entirely reasonable.

The economic analysis is likely to get interesting. A knee-jerk reaction from some economists could be to oppose short timescales. Focusing on a single fossil fuel facility, such as a coal-fired power station, the economic numbers, taken in isolation, would indicate fighting for a slow transition. However broader economic analysis, which includes the growth in renewable energy alternatives and the expansion of the new economy not reliant on fossil fuel, indicates that a short timeframe could provide a huge economic boost – despite a wave of bankruptcies amongst existing businesses within the current fossil fuel economy. In any case, the logical course of action, and only acceptable course of action, is to close down coal and unconventional fossil fuel extraction. The required course of action is clear and obvious with the only decision required over timescale.

In my view, the sooner the better. It is all too easy for the decision to be skewed by considering the existing businesses that will go bust and the jobs that will be lost. This is a blinkered, defensive approach. A bold aggressive approach is to support the new businesses and the jobs they will generate. A lost job has a face of a person no longer employed; for a new job the story of who is employed is yet to be written. It takes a leap of faith to leave a job you have, for a promise of a job which does not yet exist. As a society this is what we must do.

There are likely to be detailed discussions and negotiations over timescale which would be impossible to second-guess before even the COFF has been established. Countries least able to move quickly may seek assistance. Poor developing countries might demand cash from industrialized countries responsible for historic emissions. The much more interesting form of assistance would be the transfer of technology from rich countries to poorer countries. This need not be leading-edge high technology on which the competitiveness of rich countries depends, but relatively low technology which can be manufactured locally and installed by local labour. In this way poor countries can be enabled to help themselves, more effectively than cash transfers which can be siphoned off via corruption. There will be many countries who find they have the wherewithal and vision to prosper in the new economy, and decide to force the pace. It might be surprising which countries rise up within the reshaped economic landscape. These will not be the countries which resolutely defend the old fossil fuel economy; their misplaced efforts could hit them hard as they are left behind.

There will be some countries which resist international agreement, either for selfish reasons or because their capacity to transform is limited. However, once there is agreement amongst a significantly large group of countries, the pressure to fall into line will increase. The political and commercial incentives to develop and deploy clean technology will kill the old fossil fuel infrastructure. The existence of agreement will completely change the dynamics. Instead of the situation we have now where there are mild signals to change direction, there will be firm guidance as to what to do. Currently, we have an amber warning light telling investors to be cautious. When there is international agreement, warning lights will turn red, with firm targets and deadlines to steer investment decisions away from fossil fuel extraction.

It is too far in advance of a complex political and economic debate to provide detailed possible outcomes, but the logic is crystal clear. There are therefore grounds for optimism that an agreement can be reached with targets and deadlines appropriate to dealing with the climate emergency. As such an agreement takes shape, it would be entirely appropriate that incentives and advantages accrue to countries and corporations who lead the transition. First mover advantage will apply in any case, but it should not be beyond the wit of negotiators to incentivize those who lead the way. Countries which lead can expect to establish world leading capabilities which set them up to prosper in the new economy. Having the foresight to see beyond the short-term economic hit, as they extract themselves from fossil fuel dependency, will place the countries and corporations which adopt a leadership role in a strong position.

## Driving forward

Setting up the COFF requires action to establish it. I suggest that the most pragmatic way forward would be to raise it at a scheduled G20 meeting. This may not have the reach or authority of the UN, but the G20 is representative enough to have influence over world affairs. The host of the meeting sets the agenda. So, if the host country were to include climate change on the agenda there would be pressure to make progress. Discussions would likely go over the same old familiar ground, noting little progress. Under pressure to do better, a discussion about taking real action could take place. The beauty of the establishment of the COFF is that it is a simple decision about the way forward which need not be controversial – or at least not at the start. It would set off a process likely to lead to real outcomes. Those making the decision would need to be aware that agreeing to set up the COFF implies agreement to the concept that there should be a fossil fuel budget. It would also be fairly obvious to G20 delegates that a likely first recommendation would be to ban all new investment in coal. This should not be a surprise because this is already seen widely as necessary. A hurdle to overcome may be countries currently highly dependent on coal who will have to be given reassurance that measures will be discussed to ensure their energy security through transition arrangements. If the G20 could be persuaded, this could then be taken up by the UN, with a small cohort of countries taking the lead and setting the pace.

I predict that if climate change is regarded as an emergency, at least as dangerous over the long term as the ozone hole, that within one year of setting up the COFF it would be possible to have agreed the cap on a fossil fuel budget for the world. This is the simple maths of the fossil fuel equation. How to keep within the cap will take longer, but whereas the UNFCCC process looked to the politically impossible task of persuading countries to voluntarily limit their emissions, the COFF would focus on limiting extraction of fossil fuel. This is less controversial, and when you examine the details it is eminently possible.

I will now stray beyond that which might be predictable, into my opinion as to how discussions might proceed. As draft agreement emerges, I would suggest that it is not watered down to keep countries on board. We know from the work of the UNFCCC that weakening agreement to seek consensus is next to useless. The countries which lead the initiative should engage with drafting an agreement which suits them. Countries who remain on the periphery should not be surprised if the agreement takes shape without taking into account their national circumstances. Once a strong cohort of countries are firmly in the driving seat, it can be expected that ways are found to not only level the playing field with countries which choose to remain outside, but to go further and actively seek rules which favour countries who express the intention to ratify the convention. Ratifying the convention will be the way to gain influence on how it might evolve. In this manner, I would hope a strong agreement can emerge, gathering momentum which leads to widespread ratification. Perhaps I am being optimistic; I am certainly guilty of wishful thinking, but once climate change is seen as an emergency it could well play out in the way outlined here.

People directly involved in the COFF should be experts in fossil fuel and the fossil fuel economy. I would expect organizations like the big oil corporations to be represented and take an active role. Such powerful corporate players can be seen as opponents of climate change policy. That would be unfair. What they seek is clarity in order to make strategic decisions. When brought inside a pragmatic process I believe they could be persuaded to engage with how to manage the transition. It makes total sense for corporations within the fossil fuel industry to contribute, even if their focus is on how to profit from the transition. When the world makes it clear that there will be an end to fossil fuel, such corporations will want to work out how to cope. Currently big oil corporations do not have a clear view of the direction of travel. We as a society are not clear regarding our intentions. We talk about the need to lower emissions but then demand continued supply of affordable oil. When there is a clear political framework to close down fossil fuel, the corporations involved will make strategic plans to respond. I interviewed a director of one of the leading oil companies and asked about their approach to tar-sand exploitation. I learnt a lesson from his response. He explained that he, and the corporation he represented, could see that tar sands should not be exploited (because of the climate consequences). He went on to explain that if society allows exploitation of oil sands it would be better that a responsible corporation (like his employer) runs the operation in a tightly controlled manner minimizing environmental damage rather than less reputable corporations with lower standards. Too often we castigate oil corporations as the bad guys when they are simply doing what we as a society have decided to allow. COFF will need the commercial expertise of oil corporations to design policy for the transition which can deliver the outcome of closing down fossil fuel.

COFF will need to set the total budget for future fossil fuel exploitation, controls and timescales. That should be the COFF's work nailed down. The COFF directorate could be small and agile, with progress to agreement expected to be swift. Travelling light with little baggage is the way to press forward. A comparison with mountain climbing helps to make the point. The world's greatest mountains have been conquered by large teams backed up by huge logistics operations. They have also been climbed by small teams of dedicated and determined experts. The latter requires huge commitment and willingness to take risks; whereas the large teams can get so consumed in keeping themselves fed and safeguarded that they get stranded on the lower slopes. A lean fast-moving COFF, headed by an exceptional leader, and taking the risk of leaving some countries behind, is the way to success. This does not mean that broader issues are ignored but issues not directly relevant can be dealt with outside the COFF.

# Conclusions

Establishing a convention on fossil fuel would be a turning point in addressing climate change. Such action acknowledges that to face up to the challenge, society and the economy has to be shifted away from reliance on coal and other fossil fuels.

Rapid progress would be possible, provided there is a sense of urgency which translates into political will. Agreements to close down the dirtiest fossil fuels will not need universal support nor unanimous agreement, just a strong cohort of countries willing to lead. I hope that most reputable countries agree action for the common good, accepting that restrictions on exploiting fossil fuel are necessary. Other countries may join with the process because they see new sources of advantage which they can exploit. It will not matter that their motives are driven by national self-interest to want to become leaders of the new global economy. Seeing the future, planning to do well in the future envisioned, and then making it so, is the stuff of real politics. This would be making progress where idealism may fail to deliver.

When the COFF has completed its main work, it will have an ongoing role in managing the tail end of fossil fuel. Attention will turn to ensuring that the countries which sign up to COFF reap the rewards. The trading rules for clean fossil fuels will be a primary area of concern. Convention members will want to ensure that their economies have guaranteed access through the transition period. This will be a crucial component of making an ambitious agreement acceptable to all countries. This is where I turn next.

*The market for transition fuels*

# CHAPTER 8

# TRADING FOSSIL FUEL

Countries will need to be persuaded to agree to tough restrictions on the less efficient and dirtier fossil fuels. To allay concerns about energy security, there will need to be open trading arrangements so that countries have access to cleaner fuels through the transition period. As demand for coal collapses, strong demand should be expected for gas and oil to keep the energy system operating. These cleaner fuels will remain in demand until renewables can be ramped up sufficiently together with innovative energy storage and management arrangements to ensure reliable supply.

Once there is an agreed fossil fuel budget, and agreement that only the cleaner fuels will be used through the transition, one consequence could be a tight market for oil and gas. Oil prices, and particularly the price of natural gas, which is the cleanest fuel, could surge. There could be shortages. It will depend on how well the world faces up to the end of fossil fuel as to whether these shortages are severe and last a long time.

If governments and corporations carry on without making significant change to how energy is used, demand for fossil fuel could remain strong. The simplest response by governments, if they are thinking short term, would be to replace coal in the energy mix with gas. A massive dash for gas would put enormous pressure on gas supplies and the price could escalate. I believe that if there is a robust international agreement, governments will see advantage in taking the long view and put a strong focus on renewable energy. As governments consider long-term policy, and investors focus on secure and stable returns, it will be understood that gas is only a transition fuel. Unveiling the policy should be sufficient to incentivize the fossil fuel corporations to ramp up gas supply to serve a short-term surge in demand. With clear and unambiguous policy steering investment into renewable energy, gas supply and demand could stay in balance throughout the transition with market forces having a positive role.

'Short-term' in the context of reconfiguring energy supply systems could be about a decade with the transition complete by 2030. To many people, this statement will seem like wishful thinking, which of course it is. I want 2030 to be the target to have completed the main elements of our response to the climate emergency. Where such a bold target has been presented mainstream policymakers have not embraced it.[40] Their reluctance is not because it cannot be done but because they do not believe they have political manoeuvre for such bold action. But action is possible.[41] This is where all of us can speak out to tell our politicians to get a grip and get on with it.

One of the main benefits of a robust convention on fossil fuel is that it provides a secure regulatory framework within which corporations can make their strategic choices with confidence. Gas is likely to be in strong demand for many years with the price holding up. The tail end of oil could come sooner as the take-up of alternatives (such as electric cars) gathers pace. As opposition to fossil fuel ramps up, driven by concerns such as pollution from petrol and diesel cars, it is possible that demand for oil could reduce below production capacity. This would lead to a slump in the oil price. Oil fields which are more expensive to operate could

---

40  Centre for Alternative Technology (2010), *Zero Carbon Britain 2030*, M. Kemp and J. Wexler (eds), UK: CAT Publications.

41  Centre for Alternative Technology (2017), *Zero Carbon Britain: Making it happen*, C. Toms, A. Hooker-Stroud and A. Shepherd, UK: CAT Publications.

close down early. In particular, oil fields with heavy crude oil which is expensive to extract (in terms of both energy and cost) would become uneconomic to operate and be the first to close. As the transition gathers pace, older mature oil fields could be closed down before the reserves are depleted. The blend of government legislation and weak demand will be a killer combination.

Where a picture could be painted of the world doggedly staying loyal to fossil fuel, this is now looking less and less likely. Such a retrograde future would require extreme resistance to the forces of change. There will be resistance from some people, some corporations, and some countries. However, when the direction of travel is clear, robust controls for dirty fossil fuels in place, and alternative ways to manage society and the economy are shown not only to work but to deliver better outcomes for people and communities, resistance will evaporate. When it is clear that defence of the status quo is futile, leading to bankruptcy for businesses and pariah nation status for countries, the transition will gather pace and accelerate. Having created such a policy framework, the way to a fossil-fuel-free future would be clear of obstructions. Currently, the way forward is blocked by insisting that fossil fuel is okay as long as we agree overall carbon limits. Once the target is shifted away from carbon to focus on closing down fossil fuel, the way forward is simplified. Investment capital will flow into alternatives and business will have the incentive to lead society towards a fossil-fuel-free future.

# Lessons from the 1970s

Today the world is awash with fossil fuel and shortages are not a worry. This was not the case in the 1970s when war in the Middle East put supplies at risk and oil prices soared. The 1970s oil shock was caused by the Yom Kippur War between the Arab nations and Israel. The war began in 1973 on the Jewish holy day of Yom Kippur when Egyptian and Syrian forces entered the Sinai Peninsula and the Golan Heights. The international community got involved, with Western nations supporting Israel and the Soviet Union supporting the Arab coalition. Three weeks later the war was over with Israel having advanced further towards Cairo and Damascus, extending Israeli control in the region. The Arab nations were humiliated by their defeat and angry with the countries that had supported Israel. This was behind their decision to implement an oil embargo. The effect was a dramatic jump in the price of crude oil from about US$3.00 per barrel in 1972 to US$12.00 in 1974. This quadrupling of the oil price made some governments reflect on their reliance on oil and started initiatives to reduce oil dependency. Investment started to flow into other sources of energy and greater efficiency. An example was the open-rotor jet engine to make aircraft less thirsty. General Electric (GE) designed such an engine but it did not go into production. The prime reason was because by the time the design was ready, oil prices had stabilized and there was less commercial incentive to strive for fuel efficiency.

The start of a drive to wean the world off oil in the 1970s was sparked by political events and fears of threats to continuation of supply. At the time, the environmental movement was gathering support but a strong focus on climate change had yet to emerge. As the political crisis receded, and the environmental movement focused elsewhere, what could have been the beginning of curing fossil fuel addiction stalled. The Arab nations, and Saudi Arabia in particular, saw how in the 1970s the world considered curing fossil fuel dependency and could see their livelihood as oil producers put at risk. From then until now, the major oil producers have been careful to increase supply when oil prices rise to ensure that there is little economic incentive to reduce dependence on oil.

In my view, the 1970s oil crisis was squandered. There was an opportunity for politicians and environmentalists to come together and set off along the path to an economy not reliant on oil. Unfortunately, the timing was not yet right for such a bold move. The half century between then and now could have been ample time to complete the transition. We could now be in a much better place. The lesson we can learn is not to let the current climate crisis go to waste but to use it to drive the transition forward.

# Don't waste the climate crisis

Since the 1970s, half a century of potential progress has been wasted but we had an excuse; we did not understand the significance and danger of burning fossil fuel. There is now no excuse as the brutal reality of our addiction starts to hit home. The transition away from fossil fuel should not be feared and delayed as it was in the 1970s. The transition should be allowed to gather pace, accepting without hesitation the short-term economic consequences, with confidence that it is the correct course of action and leads to a much better place. Will the world economy suffer cold turkey? Probably. Does that matter? It should not. Will it be a negative outcome for society? That depends on whether those most addicted to fossil fuel are allowed to drive the agenda or whether those who can see beyond the addiction stand up and exert control.

It is interesting to consider how the Organization of the Petroleum Exporting Countries (OPEC) would respond to robust controls on fossil fuel. Members of OPEC have done well from oil and will continue to draw income from oil for as long as there are willing buyers. They can see the end game so countries like the United Arab Emirates (UAE) are investing in infrastructure and facilities to prepare for the future beyond oil. This includes investing in Dubai as a major holiday destination and building the city of Masdar as a 'greenprint' for sustainable cities of the future. The end of substantial oil revenues for OPEC could come sooner than expected if developments such as the electrification of transport reduces demand, leading to a glut of oil and low prices. OPEC members may not want to shoot the golden goose, whilst there is still life left in it, but they will have to respond. As the world reins back on demand, they may decide to restrict supply to hold up oil prices. The future livelihood of OPEC countries is at risk and full of uncertainty.

OPEC could respond to their concerns over future prosperity by opposing the establishment of COFF. This would be short-sighted. OPEC can see the direction of travel and know that their well-oiled cash machine cannot pay out for ever. A more astute response would be to embrace the idea of closing down fossil fuel with the aim of maximizing their return from the tail end of an industry in terminal decline.

The way to ensure that setting up the COFF can proceed without being blocked, is to make it acceptable to most stakeholders. This means pitching the proposal at pragmatists in the centre ground. There seems little point in wasting time, effort and resources in a fight between entrenched elements of the fossil fuel industry and environmental idealists making impossible demands. The strength of the proposals in this book is that they represent robust and rapid action, with something positive for everyone. There will be negatives of course, as there have to be. An analogy which is perhaps over-used, but applies well here, is that you can't make an omelette without breaking a few eggs. We have a big omelette to cook; we will need to break a lot of eggs. When people can see that they will get a share of the omelette, their attention is diverted away from the breaking of eggs.

When OPEC see that there is a share of the benefits for them, they will back policy which at first sight seems to break the back of their business.

# Open world market for oil and gas

Ensuring that there is an open market for oil and gas will go some way to alleviating short-term concerns at the rapid elimination of coal consumption. First, and most obviously, it will enable a smooth transition. Second, and more importantly, the prospect of a robust agreement to hold open markets for the clean fossil fuels will make it easier for countries to sign up to the COFF as their energy security would be assured.

The countries who will have most concern and most political difficulty in joining the COFF will be those reliant on dirty fuels such as coal and heavy oil. If such a country does not have significant reserves of gas or light oil to draw upon, abandoning coal could be seen as a risk to energy security. Poland is an example. It has huge coal reserves which it uses for generating electricity. For Poland to close coal mines and shut down coal-fired power stations will meet resistance from a number of quarters, ranging from coal miners worried about jobs to energy consumers worried at the prospect of higher energy bills. If this also means increased reliance on imports of gas, and access to gas is uncertain, Poland may resist the requirements imposed by COFF. Ideally Poland would embark on a leap forward into reliance on renewable energy sources, but such a leap would be hard without secure access to cleaner fossil fuel through the transition. A guarantee that there will be an open world market for gas will provide reassurance that an early and rapid move away from coal is feasible.

Whether an international agreement or convention for trading fossil fuel is needed will become apparent in due course. A key benefit of specific rules for trading fossil fuel is how this is can be closely coupled with the COFF rulebook. This has the potential to be an attractive carrot to persuade more and more countries to agree controls on the extraction of the least efficient fossil fuels.

The precise commitments agreed for trading fossil fuel will need to be negotiated. Those countries which sign up to COFF will be included in the open-market agreement for the clean fossil fuels. Remember that under COFF, extraction of dirty fossil fuels will be closed down. It could be that increased demand for gas during the transition outstrips supply. Countries which are signatories to COFF will get preferential access to gas supplies with all members of COFF treated equally. This will not stop gas prices overall from rising but it will guarantee continued access to supplies at a price on a par with other countries.

Considering further the idea of an open market in oil and gas, we need to be aware of possible perverse outcomes. Let us consider a country without its own oil and gas and currently highly dependent on coal. Closing coal would be a real threat to this country's economy. Let us suppose that the country therefore refuses to ratify COFF and remains outside the convention. Without guaranteed access to clean transition fuels it might double down on coal and become a dirty isolationist economy. In a world where most countries are making the transition away from dirty fossil fuels, such a country would attract pariah status and be likely to find itself on a downward spiral towards a failed economy. Remaining outside COFF will get increasingly difficult. The likely trend is that membership of COFF grows, especially if signatories to COFF accelerate the process. One way would be to agree support mechanisms and aid packages for countries most badly affected and less able to afford the required investment in alternatives. The combination of the prospect of becoming an isolated and declining economy, together with the offer of assistance, should be enough to persuade countries to join COFF without delay and accept its restrictions. It will become less and less attractive to stay outside COFF when its members are reaping the rewards. It would be reasonable that COFF members act to help each other and penalize any country that stubbornly stays outside.

# Conclusions

The transition to a world beyond the era of fossil fuel needs to start without further delay and complete before levels of $CO_2$ in the atmosphere reach levels where climate change becomes a dangerous threat. With appropriate policy, investment can flow into energy efficiency and energy generation from sources other than fossil fuel. However, this cannot be immediate; there will have to be a transition period. During the transition, bold action will be needed to complete the transformation before the fossil fuel budget has been used up. Acting as if we are facing a severe crisis – which of course we are – is the behaviour required. It is clear that to make best use of the fossil fuel budget the cleanest (most efficient) fossil fuels should be utilized and rapid action taken to eliminate the relatively dirtier fuels. We will still be living with a fossil fuel economy for some years but its nature could change quickly. Instead of the situation we have now, where there is ample fossil fuel, by choosing to use only existing reserves of gas and oil we will have designed a restricted market. How the global market for cleaner fossil fuels is managed is important to persuading countries to adopt ambitious targets, not only to eliminate coal, but also to close off investment in the energy-inefficient processes of extraction from shale rock and oil sands.

Guaranteed open access to gas and oil supplies would give countries confidence that supplies are secure and so the transition can be made. The desired end-point is the elimination of dependence on fossil fuel, so the most secure solution is to leap forward to renewable energy, reducing dependence on the dwindling supplies of oil and gas as soon as possible. The more that politicians and world leaders understand this, the more demand will weaken. One scenario is that demand does not decline fast enough, leading to escalating gas and oil prices and economic hardship. As governments and business take on board the implications of this scenario, they will behave to ensure it will not play out. The more likely scenario is that demand will scale back faster than supplies of gas and oil are depleted. It is therefore feasible that prices remain affordable throughout the transition if the COFF is backed by a robust fossil fuel trading agreement. This will deliver a transparent and predictable investment environment. There is every reason to suppose that the transition could be smooth, although it is certain to be challenging; and there will be winners and losers. A coalition of the willing operating in an open market can ensure that this is an orderly transition to the new economy.

In a decade or so, as we approach the tail end of fossil fuel, we might expect to consume relatively small quantities of gas and oil delivered from the existing ageing infrastructure. This subdued market will limp along, mainly in less advanced countries, until the final squeeze on fossil fuel when it is banned. As I look forward, I might be accused of over-optimism; but in two decades from now, looking back, we will wonder why it took so long for the penny to drop.

I do not expect my analysis to be accepted without deep interrogation, as my approach conflicts with the current dominant view. The mental block seems to be a deep-seated belief in trading carbon. In the eyes of carbon-market proponents, all that is required is to establish a global market for carbon, and then leave it to market forces to deliver the solution. Although within a perfect economic model, this can be shown to work, a pragmatic analysis shows that in the complex imperfect real world of politics, power and vested interests, carbon trading will not deliver the solution. I therefore turn to carbon markets in the next chapter, if only to show why they should be avoided.

*Carbon poker*

# CHAPTER 9

# CARBON MARKETS

Carbon markets have been put forward as a mechanism to deal with climate change, and have attracted a good deal of support. This is dangerous because whilst the belief holds that a carbon market could work, considerable effort goes into trying to establish carbon trading. This seems more like investing in a game of poker as a means to avoid reality than a real effort to come up with a solution.

We need to stop gambling with the future hoping that the carbon market can rescue us from our folly and redirect our efforts to measures with a real chance of working. This chapter is about carbon markets, simply to explain why they are a distraction and only useful in clearly defined circumstances. They can have a useful role within the tight control of states or economic partnerships such as the European Union, but this is the exception. In general, we should steer clear of carbon markets and the dangerous implication that our addiction to fossil fuel is fine as long as we trade the carbon.

The concept of global trading in carbon has a lot of followers, so it cannot be written off without explanation. I will explain both the case for carbon markets and the case against. I will also outline the circumstances where they can be useful.

If carbon markets are not your interest, you can skip this chapter. If you are curious to know why carbon markets have gained such strong support, and want to understand why, for global climate solutions, they should be avoided, continue to read on.

# An economist's view

Carbon markets arise from an economist's view of the world. From that perspective they make sense. The trouble is that economics is a step removed from reality, and in this case is a huge stride away from the real world of international politics. Carbon markets fit neatly into a theoretical simplified model of the fossil fuel economy. When you do the analysis, you discover that they reinforce the current economy rather than seek to change it. When overlaid on the fossil fuel economy, a carbon market provides a mechanism which can keep it running smoothly without disruption for many decades into the future. Unless carbon markets are designed carefully with rigorous enforcement, and with the endgame of closing down fossil fuel in mind, they can do the reverse, reinforcing and perpetuating reliance on fossil fuel. For an insight into the blinkered view of advocates for carbon markets, I will relate two incidents which give an insight into the thinking which was being employed when considering establishing carbon markets.

The first incident was at an economics conference in 2008. Leading up to the conference, I had already examined carefully the potential of carbon markets and concluded that they offered little of substance to support making real progress towards addressing climate change. I remained open to further discussion and other points of view. I wanted to find out more, to test if I might have missed something important and reached the wrong conclusion. I was keen to hear a particular economist speak who had helped to design international carbon trading mechanisms. I wanted to hear their explanation so that I would fully understand the argument in favour of a world carbon market.

The conference chair introduced the speaker with a statement that carbon markets were the key to solving the climate challenge and therefore this would be an important presentation. So, I listened attentively. The presentation was all about the economic theory of how the market could operate. There was next to nothing about the consequences of implementing a carbon market. The whole presentation was about the theoretical process with little about the pragmatic issue of how it would interact with the real world. There was what I can only describe as a 'belief' that the existence of a carbon market would solve the problem. The audience was supportive and questions were about the process and not about consequences. It was as if no one wanted to question the belief. Within the room there was solid and unwavering approval that such a market would lead automatically to the solution.

I found it quite scary to hear directly from someone designing international carbon markets that such superficial thinking had been applied. Where is the healthy scepticism to test the concept and work out if it has potential and could be delivered in the real-world context? I thought about standing up and asking why there was such a focus on carbon markets distracting attention away from the real business of bringing the fossil fuel economy under control. Sitting amongst an audience of true believers I feared my question would have been given short shrift. I took the easy option and after the presentation approached the speaker to ask a direct question. I asked, 'Have you or your colleagues thought through the endgame of how such a market plays out over the long term?'. 'We haven't considered that', was the reply. I was astonished; but on reflection I should not have been. Economists place a strong emphasis on short-term outcomes. Long-term consequences are, using the terminology of economics, 'discounted'. They are often discounted to such an extent that long-term consequences have little influence over economic policy.

This economist, and architect of carbon trading, was confident the mechanism was sound. There is a significant group of people who believe that the existence of a carbon market could magically deliver a solution – even though they have not considered the likely endgame. Not my concern, the economist might argue. But the endgame is very much *our* concern; and without an endgame, which closes the fossil fuel economy, carbon markets have no long-term purpose.

The second incident reinforced my view that economic analysis, on its own, is insufficient. It also makes the point that economists should be humble in understanding their limitations. The incident occurred at the World Forum on Enterprise and Environment 2009 held in Oxford, UK. One of the key themes of the conference was climate change and what should be done in response. At a breakout session, a small group of us discussed the role of carbon markets. Again, I was keen to discover the circumstances when trading carbon can have a positive and useful role. My approach to such markets has always been to look past the market to understand how they might contribute to dealing with climate change. This seems to me to be a fundamental and obvious requirement. A leading academic economist, who self-identified as an environmentalist, explained how a global carbon market could benefit the UK and particular the City of London if it were to become a main trading hub for carbon. That was all well and good in terms of selling the concept to financiers, but those designing the market need to have in mind the purpose of the market. Ultimately, the underlying purpose should be to address climate change. Through discussion I probed further. This economist was not interested in long-term outcomes. Despite having a role in advising government about how such markets could operate, this economist seemed to have made no attempt to understand how the market would play out in the long term. Once again, the belief that the existence of the market was sufficient was dominating the thought process. This was someone who was genuine in believing that setting up a carbon market was a key step to solving climate change, even though they had not critically examined whether it really could.

The economists mentioned in these incidents shall remain nameless. It could have been any number of economists and I feel that it would be unfair to single out individuals. The problem is simply that economics alone is not enough. If carbon markets exist, they will be played for profit; but the people designing the market have to be crystal clear what needs to be achieved. Without a clear driving purpose, all they are doing is setting up a casino which gives the impression of being helpful, when in fact it is a distraction from real policy to deliver positive outcomes.

The way that concepts derived from economic analysis become belief frameworks is all too common. The economists who propose carbon markets have sound economic arguments to support their case. The recommendation is not questioned but accepted as a core belief, often by people who are not themselves economists. Politicians are particularly likely to fall into this trap. The danger is that a blinkered approach which accepts the economic case without question can lead to poor decisions or inappropriate policy. The only safe way to proceed is to examine the logic of the particular case. The concept of carbon markets has been elevated to become a belief framework; this has to be challenged.

Belief in carbon markets is ill-founded, but the problem with belief is that it sticks in people's minds and cannot be shifted easily. When an open debate takes place about an issue, factors are discussed and options considered. When a belief framework is challenged, those holding the belief will resist balanced argument and oppose alternative views until a cast-iron case can be made, backed up by concrete examples. This chapter challenges the belief that carbon markets can solve climate change. I hope I present a cast-iron case, so that policymakers can now move on past this particular distraction to considering action with the potential to make real progress.

# The case for global carbon markets

To challenge a belief, it is often better to begin with examining the belief itself and define clearly the argument which supports it. Once this is recorded, it provides a solid base from which to deconstruct the logic. Those who hold the belief are then forced to question why they believe what they do. This is how I will approach the examination of the concept of global carbon markets. In order to deflect accusations that my opposition to carbon markets is one-sided, I will start by stating that they do have potential. If you understand how they operate and consider long-term consequences – and use these insights to design appropriate regulations and controls – then carbon markets can indeed have a positive role.

To begin the case in favour of carbon markets, let us define the elements of the market. The three elements are a cap, permits, and a trading mechanism. The cap is the total amount of $CO_2$ that can be emitted from all sources. The cap should be set according to what policymakers, working with advice from climate scientists, decide is an appropriate upper limit. The permits are permission to emit an amount of $CO_2$. The permit conveys the right to burn fossil fuel from any source for any purpose. Businesses emitting $CO_2$ need a corresponding permit. The key to how the market works is that the total number of permits is capped. Permits can find their way into the market in a number of ways. They can be handed out based on records of past emissions, sold or auctioned. Anyone wanting to emit $CO_2$ must hold a permit. The trading of permits allows the market to set a price for the permit, which in this case is the price of carbon. Assuming that the market is tightly regulated to operate as intended, total $CO_2$ emissions will be held within the cap. So that is the theory.

Players in the carbon market can do a number of things. $CO_2$ emitters can save on the cost of purchasing permits by reducing their emissions. The price of carbon becomes the yardstick for investment decisions with regard to energy efficiency. There is uncertainty of course because the price of permits is set by the market. If considerable investment flows into energy efficiency reducing the demand for permits, the price of carbon goes down. The consequence is that energy efficiency becomes less financially attractive. A dynamic balance is struck between investment flowing to keep emissions beneath the cap and investment withheld to steer emissions to rise to meet the cap. Reducing emissions faster than complying with the cap is penalized. Instead of being an upper limit to emissions, the cap tends to become a target to meet.

Other players in the carbon market can propose projects which strip $CO_2$ out of the atmosphere, such as forest planting. These projects can claim permits to sell on the market. So, the market sets a price for carbon and steers investment decisions towards using less fossil fuel and stripping $CO_2$ out of the atmosphere.

If you reflect, it is not the market which is of primary importance but the price of carbon. Using the market mechanism means the carbon price is variable. Rather than drive carbon out of the economy, it would tend to find the equilibrium which would hold carbon consumption at a predefined level. I suggest that this should not be the purpose of the market. The purpose should be to squeeze carbon emissions lower, and maintain downward pressure on fossil fuel consumption. For this purpose, a fixed price for carbon would be better, especially if it was a predictable price; and perhaps a rising price escalator published well in advance. I want to focus on the case in favour of carbon markets, so let me leave this troublesome observation to one side for now.

A global carbon market could in theory work, and is attractive because it doesn't require much of countries and corporations. Buying and trading permits is politically much easier than closing down particular fossil fuels. The true test of a carbon market is whether it delivers the reduction in emissions required. In theory it could, provided the cap was reduced in line with a downward trajectory which fitted the total emissions to avoid dangerous climate change. This is not just a side issue; this is the key issue often conveniently ignored. The cap on emissions would need to be effective and reduce over time. For a global carbon market, the cap would need to be set for the world as a whole with enforcement mechanisms which apply universally to all countries. This would have to be brokered through international agreement. There are some who believe that the UNFCCC could deliver such an agreement.

With its focus on carbon, the market is neutral over which fossil fuels are displaced, which processes are altered, and what infrastructure is improved or replaced. It is assumed that the invisible hand of the market will steer investment to where it will have the most impact. Rather than central planners and policymakers having to take direct control, they can trust the market to deliver.

The theoretical economic case for carbon markets is sound but questions arise, particularly with regard to designing one which operates at world level. Could a total cap which reduces overtime be agreed? Could such a cap be enforced? When prices of permits escalate as the cap is squeezed tighter, would the agreement hold? It does not need much reflection to understand that a political crunch point will be reached which breaks the market. It could be that with huge effort a global cap is agreed and a worldwide mechanism of enforcement established. However, that would just be the start, delaying the crunch point. The political crunch point at which demand exceeds the reducing cap is almost inevitable.

So, I have provided a summary of the case for a global carbon market apparently based on rock-solid economic theory. However, it is already apparent that the requirement to agree and enforce the cap is its Achilles heel. Having sown the seeds of uncertainty that the global carbon market may not be as good as was thought, I will now go further and lay down the logic to explain that such a global market cannot work.

# The case against global carbon markets

The core purpose of establishing a global carbon market should be to deal effectively with climate change. There may be fringe benefits but unless the market can deliver on its core purpose there is little point in expending effort and resources in setting it up. Building a casino, fronted by good intentions, backed up by weak controls, is not what is required. Such a casino market will be exploited by those seeking to profit without delivering benefit back to solving the climate challenge.

Let us begin examination of the case against a world carbon market by considering what is inside and outside the market. All fossil fuel consumption would need to come under the umbrella of such a market: coal, gas, oil, fracked gas, oil from tar sands, and any other source of fossil carbon extracted and released. In addition to requiring that all sources of fossil fuel are caught within the market, there needs to be clarity as to who is required to hold permits. Does the permit apply at the point of extraction or at the point where the fossil fuel is burnt? There will be considerable discussion about who pays for a permit but let us assume that a system could be agreed and established. In a world carbon market, rules and regulations would need to cover all countries with a tight, legally binding and transparent system, free from fraud and corruption. You would need to be trusting and optimistic to believe that such a system could be established, but less us assume that it can be done. The case against such a market is more fundamental than difficulties of implementation.

A carbon market is not just about fossil fuel. Activities which take $CO_2$ out of the atmosphere are allowed to be part of the market. This brings us to the first problem. Allowing forestry, for example, to participate in the market seems reasonable to an economist with a short-term view. As the trees grow, they take carbon out of the atmosphere. It seems sensible to allow those who plant forests to raise money to do so by selling permits. However, when the forest matures, what happens next? If the wood is harvested and buried deep underground, the carbon would indeed be locked away, offsetting fossil carbon emissions. If the timber is incorporated into the construction of buildings, it could be locked away for many decades. Where it is used for buildings of quality, such as the roof of Notre Dame Cathedral in Paris, it could be kept out of the atmosphere for centuries. Whatever route the forestry carbon takes it will eventually be released when the timber comes to the end of its life, and presumably burnt.

Bringing forestry inside carbon markets seems like an attractive and useful idea. It can have useful consequences in preserving forest cover but we should not be fooled into thinking it helps with regard to solving long-term climate change. The operation of a world carbon market should not be deflected away from the core purpose of limiting overall carbon dioxide levels in the atmosphere. We need to look after forests, biodiversity and many aspects of the natural environment but not in ways which become excuses for our failure to control fossil fuel consumption. There are other problems with playing the forestry card in negotiations over allowable emissions. When a country offers to plant a new forest or reinstate forest which has been cleared, there is a clear net benefit to maintaining natural capital. However, the negotiating tactics for countries which are in possession of large forests can take the direction of threatening to clear them unless payment is made, or permits issued, to keep the trees standing. Countries rich in virgin forests could hold the world to ransom, claiming considerable income to retain the forest. This could play out over years or decades, until the countries being blackmailed decide enough is enough and stop paying. When the income dries up the blackmailer will be free to clear the forest in any case. Preserving and maintaining forest requires good governance and good environmental policy but let us not mix up this important environmental issue with the carbon market. It could go so terribly wrong.

The carbon market has to be focused on fossil carbon if it is to make any sensible contribution to ensuring long-term reductions to climate change. There are some useful and valuable possible ways to incentivize countries, especially poor countries, to retain virgin forests. Pretending that this could be achieved through the carbon market comes from blind belief in carbon markets without thinking through an analysis of their long-term consequences. The fundamental problem with the proposed world carbon market is that it is too far removed from the core purpose, which should be to deal with climate change, which in turn has to deal with keeping fossil carbon locked underground.

Perhaps the worst problem of a global carbon market is that it would give the impression that $CO_2$ emissions are being brought under control. The world could be persuaded to believe that climate change can be addressed by extending the carbon market to all corners of the planet. This story of progress could be a front for inaction which continues for decades. Whilst the belief in carbon markets is allowed to persist it could block measures with a greater chance of success. This is the strongest argument against investing further political capital in establishing carbon trading and a world carbon market – it is distracting effort which could be applied more effectively in other ways.

I have explained why the belief in global carbon markets is misguided but I may not have done enough to persuade true believers to change their view. We need to return to the core purpose and examine carefully how a world carbon market plays out over the long term.

Economists may prefer a short-term perspective, arguing that the long term is outside direct control. In the case of climate change, its fundamental nature is that it is long-term. The consequences are damage to the planet in the decades (and centuries) ahead. The risks are to the long-term livelihood and security of our children (and their children). With consequences this severe, and risks so great, we cannot afford to discount the future (in the way favoured by economists). It is crucial to examine the long-term consequences of a world carbon market and to use such insight to decide what should be done right now.

I will commence my examination of the long-term outcome of a world market for carbon by making some assumptions. I will assume that a global carbon market can be established; that it would have global reach; that it will be bound by tight regulations; and rigorously policed. These are assumptions which could easily be disputed. The reason I want to make and accept these assumptions is that the flaws in a global carbon market run deep. Even if the market was set up and operated as intended, it would still be a dead duck in relation to dealing with climate change.

The critical component of any global carbon market would be the total cap on emissions. This cap would be set according to the sum total of $CO_2$ emissions before climate change becomes dangerous. This is the number discussed over many years within the UNFCCC and agreed by all parties; and the same numbers that are used in this book. This is a solid and undisputed baseline with which any action to deal with climate change should conform. This sum total applies to the remaining years of fossil fuel consumption, however long or short that might be.

As explained in Chapter 6, the remaining period of fossil fuel consumption would be about 14 years if emissions are capped at current levels (to meet the 1.5-degree target). It would be shorter if emissions continue to creep higher or longer if the cap on emissions is screwed tighter, quickly conserving the allowable emissions budget. The point will come when the total cap on overall emissions has been reached and fossil fuel consumption must cease. This would be the end of the market according to the theory of how it is set up. Let us consider two key aspects. First, what has been happening whilst the market was issuing permits; and second, what happens when the market reaches its end point and is due to close.

Whilst the carbon market is running, all fossil fuels can be consumed but the overall carbon cap is enforced at an agreed annual allowance. Remember that we have assumed this is all agreed and will work as intended: gas, oil, coal, conventional oil, and fracked gas are all available. Because the market is directly accounting for carbon, the cleaner fuels which give more energy for the carbon emitted are favoured. However, decisions will be taken calculating both the cost of permits and the cost of the fuel. Where a dirty fuel (such as coal) is sufficiently cheaper so that the total cost of the fuel and permit is competitive, it will continue to be burnt. It is argued that the way the carbon market is neutral about which fuels to use is beneficial. From a price-setting prospective, this might be so; but when you understand the core purpose of the market it makes no sense whatsoever. Those familiar with markets and market forces might be starting to get confused. We need to understand that the market should not be about minimising the price of continuing to exploit fossil fuel; it has to be about managing the transition away from reliance on fossil fuel.

During the early years of the carbon market, while clean fuel such as gas is plentiful, the market will favour gas, but as gas supplies get tight the market will move smoothly and efficiently to ramp up the use of oil, shale oil and even coal. It might be hoped that the carbon cap is tightened faster than gas reserves are depleted so that a resurgence in coal does not happen, but there are no guarantees, as no attempt is made to regulate which fossil fuels are removed first from the market. That is decided by market forces. This means that as the world uses the remaining allowable emissions of fossil carbon this applies to the full range of fuels. The sum total of energy produced from the total allowable emissions will be less because we will have allowed dirty fuels to continue as part of the mix. The market has not maximized the energy delivered from the remaining allowance of $CO_2$ emitted; but minimized the price of continued reliance on fossil fuel. This will leave the world in a difficult place.

An even more fundamental problem with the global carbon market comes from examining the point at which the permissible allowance has been reached and the market should close. Up to this point the world has continued to use a range of fossil fuels, so there will still be reserves of natural gas and conventional oil. You would have to be a naïve dreamer of epic proportions to think that the world would at this point close off such supplies. The world carbon market would be relatively benign in the early years, steering consumption towards gas, but would reach a point when fossil fuel consumption should close but it will be politically impossible to do so.

Once you reflect on and understand how a global carbon market would operate, particularly in the later stages, it becomes clear the outcome is to delay tough choices without solving anything. All it would really do is to support a smooth transition from clean fossil fuels, when they are in abundance, to less efficient fuels as supplies get tighter. The initial political challenge of establishing a tightly regulated carbon market would be tough, but the political resolve required to hold the market to account in the latter stages would be much more difficult.

Global carbon markets would be worse than useless. Setting them up will expend political capital which could be better employed. They will give the false impression that emissions are being brought under control. Carbon markets will reinforce the fossil fuel economy rather than dismantle it, smoothing transition from clean to dirty fuels. It would be politically impossible in the later stages of the market to hold to the overall carbon cap. Most significant of all, the existence of a global carbon market would delay the tough choices which have to be made in any case. Each further delay makes the task more difficult.

If we really intend to face up to climate change, we have to focus on fossil fuel. We don't want to get trapped in a carbon casino designed so the ordinary punters (humanity) are bound to be losers. Fossil fuel addiction is bad; adding a carbon gambling habit would not be helpful.

## Closed carbon markets

Having outlined the problems with a global carbon market, it is worth examining whether carbon markets can have a positive role to play. My conclusion is that carbon taxes are almost always likely to work better, but for a complete analysis let us examine in what circumstances a carbon market could operate successfully.

At regional or national scale, under oversight of competent authorities, a carbon market can work. Such a market could facilitate investment flows to where energy efficiency gains can be made without government needing to provide top-down control. The detailed decisions would be left to market forces. The market cap would be set by the government in line with whatever international commitments had been made to control carbon emissions. Presumably it would be ratcheted down over time as fossil fuel is squeezed out of the economy. It would be hoped that national commitments to reduce carbon emissions would be a component of a global agreement. Even if such a global agreement was weak, well-run responsible countries could still set and hold to targets which ratcheted down over time.

The government would have to make choices about how to allocate permits. One way would be to provide business with a free allocation according to their existing emissions at the time the market commences. Another way is to sell permits at a particular price. A third way is to auction the permits so that the market sets the price. As the total amount of permits allowed each year is reduced, the price adjusts providing a dynamic balance. If carbon emissions are expensive, business will invest more to reduce them. If emissions have been dropping faster than the reducing cap, the price of permits will go down, slowing further investment. In this way the economy will adjust to squeeze out carbon in line with the cap which reduces over time.

A possible perverse outcome of the market is if significant carbon emission reductions become viable through technological breakthroughs in either energy efficiency or energy generation. This could lead to dramatic reductions, which of course would be good. The problem would be that having delegated decisions to the carbon market, the price of permits would crash so that investment in the new technology would dry up. In effect the market not only holds emissions down to remain within the cap but also perversely can hold them up to meet the cap. The ambition to make reductions is set by government rather than what would be possible by innovative companies pushing forward quickly. It is ironic that the market mechanism designed to remove government control over detailed decisions ends up exerting total government control, limiting what can be achieved.

The more you examine carbon markets, the less attractive they look. The price of permits is volatile so business plans and the justification for investment are uncertain. The market holds emissions up just as much as it holds them down. It seems odd to be bothered with carbon markets at all, but they do have one huge political benefit in that the price of permits is set by the market and not by politicians. Carbon taxes are much more precise, supporting sound business plans and giving investors more confidence, but governments get the blame for setting tax rates and governments would come under pressure to keep carbon taxes low. Politicians would have to find the courage to set carbon taxes high enough, to decarbonize fast enough, to hit the targets required.

There is another aspect of the carbon market concept which can be seized upon as beneficial. This is to allow the market to extend outside the market for which it was set up. The apparent benefit is that it allows a carbon reduction project in another jurisdiction to generate a carbon offset which could be regarded as equivalent to an additional permit to pollute. If the project really does reduce $CO_2$ emissions then, according to the logic of the argument employed, the overall impact is neutral. So, a business emitting carbon dioxide within the market can pay for a project elsewhere, instead of investing to reduce its own emissions. This might well be cheaper, particularly if the business is located in a country where carbon efficiency is already relatively good. Within a carbon market it wouldn't be necessary to connect a particular project with a particular polluter; two national carbon markets could be linked allowing carbon permits to flow from one to the other. If both markets are tightly regulated and held to account by ambitious and reducing caps then this could make sense. In the real world, allowing carbon to be traded across jurisdictions is more likely to be a recipe for corruption and fake reductions. This might be welcomed by unscrupulous businesses which are happy to pay to avoid reducing emissions and are not concerned with looking closely at where the money goes. Such 'benefit' looks more like an excuse for inaction and would be beneficial to those businesses which can't be bothered to get on with the job of decarbonizing.

I accept that, in theory, national carbon markets can be made to work. Such markets can support the transition from the fossil fuel economy to a low-carbon economy. It must be borne in mind that they are only a fig leaf unless tightly regulated and controlled. Carbon taxes are a much better tool of policy, providing clear and unambiguous guidance for investment decisions. Such taxes can support rapid progress as innovative methods and improved technologies are deployed without penalizing the companies which can orchestrate more ambitious progress than overall targets set by government.

## Conclusions

Carbon markets should only be established after careful examination of the consequences and regulated to deliver long-term sustained carbon dioxide emission reductions. In my view, they are overrated; it might be better to consign the concept to the dustbin of potentially good ideas which do not stand up to careful examination. This would free up resources and effort to investigate other ideas with a better chance of success.

As we navigate the transition, the focus should be on energy efficiency and energy generation from zero-carbon sources coupled with closing down the extraction of fossil fuel. Using market forces to put a value to $CO_2$ can undermine this, smoothing the transition from clean to dirty fossil fuels, thus entrenching fossil fuel dependency. I know this is not the intention of those who propose carbon markets, but unless the markets are rigorously policed that is the likely long-term outcome.

There are short-term attractions to using carbon as a proxy for the elimination of fossil fuel. Carbon markets can play a role in retaining and regulating the fossil fuel economy. In the early years of the market it is less painful than real measures to regulate fossil fuel. However, we don't want to retain and regulate the fossil fuel economy when what we should be doing is working on replacing it with a quite different model. If we rely on carbon markets, when these markets mature a crunch point is reached when squeezing the cap tighter (in accordance with climate policy) makes the fossil fuel economy very expensive. The political pressure to ease the cap to keep the (fossil fuel) economy on track will then become intense, which will put climate policy at risk. Pragmatic politicians may have little choice but to abandon the cap, totally negating the purpose the market was presented as solving.

Carbon markets have been tried in the EU and elsewhere. We should now know enough about their limitations to move past this distraction to get fossil fuel consumption under control. The real game is not to squeeze carbon out of the fossil fuel economy, but to replace the fossil fuel economy with a different economy that is not reliant on fossil fuel.

*Tax-free aviation fuel*

# CHAPTER 10

# AVIATION – A TEST OF RESOLVE

Grasping the thorny issue of emissions from aviation will test our resolve to actually do something about climate change.

When it comes to facing up to a difficult challenge, there are always many reasons not to. It might be too difficult, because a solution is not obvious; because there is huge opposition; or because you just can't quite conjure up the resolve to get on with it. In the end, the challenge of climate change must be faced. Yes, it will be difficult; yes, the complete solution is not obvious; yes, there will be huge opposition to effective action; and yes, we will find the resolve – eventually. There is often a pivotal moment when you break out and decide to get started. It might be the first trip to the gym as you decide to get fit; mark the first of a pile of exam scripts; or buy the paint in order to decorate the front room. For the challenge of climate change, the pivotal moment will be when we find the resolve to take action over aviation. All of us are at fault for failing to insist that aviation is reformed; even those who choose not to fly do not do enough to call for systemic change which applies to everyone.

Dealing with the climate impact of aviation is obvious – when you do the analysis – but it will have short-term negative consequences for both the economy and the travel choices available to passengers. The action required will not be liked by some countries; not liked initially by passengers (although that will change); and in particular will not be liked by the aviation industry which would suffer a wave of bankruptcies. The current tax-free exemption for aviation fuel traps the industry in the twentieth century. Remove this, and aviation will be consumed by a firestorm of innovation, making way for a new model of air services to arise out of the ashes of a burnt-out industry like a phoenix taking flight.

## An industry stalled

The huge expansion of aviation, which began in the twentieth century, continues apace but with little thought for the consequences. It is common knowledge that the environmental impact is not being tackled, but the industry remains in denial. It is flying ever higher on a steep trajectory of more routes, more planes, and more passengers. The analogy which comes to mind is of an aircraft with the pilot pulling back on the stick and climbing steeply. Unless the pilot eases back, the plane will stall and crash to earth. This is where aviation is in the early twenty-first century – climbing steeply, at risk of crashing. My fascination with this industry – which seems to have a death wish – continued as I expanded my analysis and wrote the book *Fly and be Damned*.[42] I deliberately chose the title to have a dual meaning. It can be taken to mean: you should not fly because of the environmental impact and you will be damned if you do. It could also be taken to mean: I will fly because it suits me, whatever the environmentalists say. My view is neither of these extremes; I want to fly but I want my flight to have low environmental impact. Since publication, my ideas have received support from environmentalists, passengers and politicians; but not the aviation industry. The industry either does not see, or conveniently ignores, that society will in the end demand better flying choices.

---

42  McManners, P.J. (2012), *Fly and be Damned: What now for aviation and climate change?* UK: ZED Books.

More recently, I carried out research into aviation to examine why progress remains stalled. I discovered that if only people understood how much better aviation could be, there would be widespread support for transformation to a new and different model of aviation.[43] Aviation is trapped in a highly polarized debate, with on one side environmentalists urging people not to fly. On the other, the industry is resisting change as they struggle to make profits in a highly competitive sector. Government does little, because there is no popular demand for change. There is collective failure. Environmentalists, the industry and government, are all mistaken in their approach. Environmentalists should be arguing for green aviation; meaning flying with low environmental impact. Rather than ranting against flying, environmentalists should be demanding the transformation of flying to a different model. For the aviation industry, it will have to change eventually, so executives should be planning how to navigate through the transformation. Instead they sit tight and hope that they can survive from one set of annual results to the next, without the industry crashing on their watch. The government is wrong to park the issue as just too difficult, because aviation policy can be fixed. To do so would show real resolve to face up to climate change by addressing one of the thorniest issues.

The manuscript for this book was finalised whilst the world was gripped by the COVID-19 crisis. Airline bosses can no longer sit tight and hope for the best, but must respond if they are to survive. Flights have been cancelled, aircraft grounded and a wave of bankruptcies could follow. Governments are being asked to bail out the industry but they will need some guarantee that

---

43  McManners, P.J. (2017), *Reframing Economic Policy towards Sustainability: explored through a case study into aviation*, UK: Routledge.

there is a viable industry going forward. This crunch point would have come eventually as climate policy tightened but COVID-19 has brought it forward. It would be sensible to allow creative destruction to spread through the aviation industry to allow the development of new ways to fly. If there is to be financial support, it should be used to launch new, cleaner ways to fly, not to defend an outdated industry.

It might be years before aviation demand takes off again. This is a wonderful opportunity to develop a model of low-emission air transportation. Immediate action is likely to be that all but the most modern and efficient aircraft will lose value. Older aircraft fleets may become worthless and never fly again.

Before COVID-19, the expansion of aviation was expected to continue or even accelerate. China alone expected to build 200 new airports by 2035.[44] Continental Africa has been seen as a huge untapped market yet to take off. The Indian aviation market has massive potential for growth. The International Air Transport Association (IATA) has been bullish about the prospects. In 2018 the IATA reported that present trends in air transport suggest passenger numbers could double to 8.2 billion by 2037. The IATA's Director General, Alexandre de Juniac, said:[45]

*Aviation is growing, and that is generating huge benefits for the world. A doubling of air passengers in the next 20 years could support 100 million jobs globally.*

The COVID-19 crisis has dented such optimism. IATA, in a

---

44  Cummins, N. (2019), 'China Plans to Build an Astonishing 200 New Airports in The Next 15 Years', [available online: https://simpleflying.com/china-to-build-250-new-airports/; accessed 20 May 2020].

45  IATA (2018), 'IATA Forecast Predicts 8.2 billion Air Travelers in 2037', press release 62, 24 Oct. [available online: https://www.iata.org/pressroom/pr/Pages/2018-10-24-02.aspx; accessed 5 Nov 2019].

joint statement with the Airport Council International (ACI), stated (IATA 2020):

> *Today we face the biggest challenge in commercial aviation's history: Restarting an industry that largely has ceased to operate across borders, while ensuring that it is not a meaningful vector for the spread of* COVID-19.

The industry's initial response to the pandemic was totally focused on restarting the existing model. This blinkered view is understandable in the midst of the crisis, but it is astounding that the aviation industry has over many years remained deliberately ignorant of the environmental consequences of its operation, claiming that expansion of the current model of aviation demonstrates 'huge benefits for the world'. The industry may attempt to continue to plead ignorance but when the world faces up to climate change and puts aviation under the spotlight, there will be a reckoning.

# Limiting emissions

During international negotiations to limit carbon dioxide emissions, aviation has been left outside as a special case. The International Civil Aviation Organization (ICAO) has been tasked with making its own proposals. In response the industry has published the following 'aspirational goals':[46]

+ 2% per annum fuel efficiency improvement

+ carbon-neutral growth from 2020 onwards.

At first sight this seems to be useful but it is entirely

---

46  ICAO (2019), 'Destination Green; The Next Chapter', Environment Report.

disingenuous. ICAO expect continued expansion and for emissions to ramp up considerably. The pledge is based on carbon offsetting and alternative fuels such as biofuel, arguing that this would be carbon-neutral. In October 2016, the ICAO adopted a scheme called the Carbon Offsetting and Reduction Scheme for International Aviation (CORSIA). Measures include offsets and alternative fuels but addresses only emissions that exceed the baseline of 2019–2020 levels. The industry expected 2020 emissions to be high therefore requiring the industry to do very little. The COVID-19 crisis transport lockdown could mean that a 2020 baseline becomes challenging. I would not be surprised if the rules are rewritten to ensure CORSIA has little real impact.

An interesting exercise is to consider whether it would be possible for aviation to switch to biofuel. Depending on the assumptions you make, my rough back-of-an-envelope calculation is that to switch aviation to biofuel by 2030 would require the entire agricultural capacity of all the farmland in Europe.[47] The airlines have pilot projects to make biofuel from waste but these could only be a tiny drop in the tanks of a thirsty aircraft fleet. The other aspect of their approach is to require airlines to offset their emissions by purchasing emission permits. As explained in the previous chapter, this is a weak approach and to be avoided if we are to face up to effective action. The plain truth is that there are no plans to limit emissions from aviation and seemingly no appetite within the industry to do so.

The industry is not volunteering to change, but governments could force change, if pressurized by sufficient public concern. The transport lockdown during the COVID-19 pandemic showed

---

47  Assumptions: Aviation current demand is 5 million barrels of oil per day rising to 7.5 million barrels per day by 2030; one barrel of oil contains 1.7 Mwh of energy; EU agricultural land 173 million hectares; Biofuel production at 30 Mwh per ha per annum.

people, particularly those living under flight paths, the value of clean and quiet skies. There is a window of opportunity when public indifference could switch to demands for improved flying options.

Business is capable of responding, if forced to do so. The business plans of aircraft manufacturers are based on selling current models expecting a flying life of two decades or more. When this is no longer the case, the balance sheets of manufacturers and the airlines will be hit hard. Although not welcomed by the industry, this would require writing off investment in current models. The industry has the technology for low-emission flying but not the business case to develop it because fuel is cheap. As new capital comes into aviation, investors will demand a secure sales pipeline. Provided government policy has been clarified, the stage could be set to initiate a transformation in aviation. The unexpected arrival of COVID-19 has already begun the destructive phase. The key to getting this right is to launch the recovery plan in a climate-friendly direction.

The oddity about the situation, and the reason a crash was inevitable, are the rules which govern civil aviation. The rules are set by international convention; they are tight, detailed and hard to amend. In this safety-critical industry, this is to be expected. Such a slow and ponderous system of regulation means it lacks the agility to adapt to changing circumstances. The industry remains trapped in a set of rules agreed in 1944, which have changed little since. Additions and adjustments to the rules can take years to get discussed and agreed, but rewriting the set of rules is not out of the question. For a rulebook which is now 70 years old, it is high time it was updated to take advantage of the huge potential to improve and advance aviation in the twenty-first century.

There is one particular rule in the Convention on International Civil Aviation which stymies progress. It is a rule so deeply embedded in the fabric of the industry that it is seldom questioned. I have read a number of books on aviation economics. They are not the most riveting of reads but they all assume that this rule will not change. We should not blame the economists who wrote the books; they are reflecting the current political reality. We should also not blame the industry for failing to advance, because this rule undermines the business case for bold advances in reducing the environmental impact of aviation. The industry is not going to invest if they are sure to lose their shirts. We should be blaming politicians for allowing this rule to stand without being questioned. We should also be blaming ourselves for not telling our politicians that we demand change. For most people there is a good excuse; they have never heard of the rule – and they have no idea of the significance of changing it.

We should not blame those who wrote the treaty; their intentions were good. The preamble to the treaty was written during the final stages of World War Two and illustrates their overriding concern to use aviation policy to help cement the peace after a period when the world had confronted a terrible international crisis.[48]

> *The future development of international civil aviation can greatly help to create and preserve friendship and understanding among the nations and peoples of the world ... to promote that co-operation between nations and peoples upon which the peace of the world depends.*

The world again faces a terrible international crisis; it is time to rewrite the rules of aviation policy in order to respond to the current crisis.

---

48 International Civil Aviation Organization (ICAO, 2006), 'Convention on International Civil Aviation' – doc 7300/9. Ninth Edition dated 2006, [available online: https://www.icao.int/publications/Documents/7300_cons.pdf; accessed 13 Nov 2019].

The current rules of international aviation were agreed in December 1944. This was seven months before the armistice which ended the war in Europe. The world was still in the midst of the crisis but leaders had the forethought to be thinking beyond the crisis to ensure that it would not be repeated. The convention was convened by the United States in the city of Chicago. Many of the delegates travelled at great risk from war-torn and occupied countries to attend. In little over one month the rules which were to govern international aviation were drawn up and agreed. This is largely the same set of rules which apply today. When the world comes together to tackle a crisis, such tight timescales for decision-making become possible. It makes the current long-drawn-out deliberations over how to respond to climate change look pitiful in comparison.

In many ways the Chicago convention has served the world well. In 1944, 'promoting co-operation between nations and peoples upon which the peace of the world depends' was an appropriate aspiration for the development of international aviation. Over 75 years later, the world situation has changed, the challenges are different, our aspirations for aviation should be reset. The 1944 'rulebook' required that 'international civil aviation may be developed in a safe and orderly manner ... on the basis of equality of opportunity and operated soundly and economically...' These words are fine but are no longer sufficient. We need to add: 'without negative environmental impact'. For pragmatic reasons, our aspiration could be set at: 'whilst minimizing the environmental impact'. It would be possible; we have the technology and know-how; what we don't have is the will or an international treaty which agrees that we should. The solution is for world leaders to convene a new convention with the explicit task of agreeing arrangements to maintain international air transport services whilst minimizing the environmental impact. It should be no surprise that after 75 years the rules are no longer fit for purpose. We face a climate crisis (and in 2020 a pandemic). Let us mimic the success of Chicago 1944 and convene a convention during a time of crisis to agree within one month a new rulebook for international civil aviation. Time is of the essence to do this before resources start to flow back into rebuilding aviation post-COVID-19. To use 2021 and beyond to put aviation back where it was, would be a huge waste of a golden opportunity.

Once the principle of minimizing environmental impact is established and agreed, aviation experts would be empowered to put aviation on a sustainable track for the twenty-first century. There is one particular article of the current rulebook which holds back progress and locks aviation into the fossil fuel economy. Until this is changed, there is no escape for the industry and the conflict between environmentalists and air services customers will not be resolved.

Article 24 of the Chicago convention is the culprit and block to progress:

> *Article 24: Aircraft flying to, from or across, the territory of a state shall be admitted temporarily free of duty. Fuel, oil, spare parts, regular equipment and aircraft stores retained on board are also exempted from customs duty, inspection fees or similar charges.*

This appears entirely innocuous. It seems to be bland and routine detail, until you understand the consequences. I need to walk through the logical steps to explain why Article 24 is such a problem for the future of the aviation industry:

**Step 1:** Article 24 prevents countries from taxing fuel carried on board planes landing at their airports.

**Step 2:** Textbooks of aviation economics recommend that airlines fuel up where the fuel is cheapest and avoid taking on fuel where prices are higher. Even small differences in price can matter because large jet aircraft burn such huge quantities of fuel.

**Step 3:** If any country were to decide unilaterally to tax aviation fuel, airlines would avoid filling up in that country. So, if the UK for example were to decide that it is about time that aviation fuel was brought more in line with fuel for ground transportation, and taxed, it would be an economic own goal. Instead of Heathrow being a major international hub for international air transportation, the airlines would relegate it to a regional destination. This would be a normal commercial decision by airlines to remain competitive.

**Step 4:** The consequence of steps 1, 2 and 3 is that governments do not tax aviation fuel for international flights, and any discussion that they might do so is soon closed down.

**Step 5:** Although from time to time this fuel tax anomaly comes up for discussion, tax-free aviation fuel is hardwired into the aviation economic model. The reason for this is that the only way countries will be persuaded to tax aviation fuel would be through joint international action; and there is no appetite to do so. The United States in particular is ruthlessly opposed to such a tax.

**Step 6:** Although in theory countries could break ranks to tax aviation fuel, they will not unless Article 24 is changed. The United States took the lead in drafting the 1944 convention and retains what is effectively a veto over the process of amending the convention. These six logical steps mean that aviation fuel for international flights is tax-free; and everyone in the industry, policymakers in government, and even environmentalists accept without question that this will forever remain the situation.

The consequences of an economic model for aviation based on the principle of tax-free fuel are significant. First, where governments seek to use taxation to restrain emissions from flying, the tax falls on passengers, such as the UK's Air Passenger Duty (APD). This is politically difficult because taxing passengers is seen as unfair to those less able to afford the higher prices. Governments therefore are not going to put significant taxes on flying. The more fundamental problem is that taxing flights might slow the expansion of flying within the current economic model, but provides no incentive to switch to a low-carbon economic model.

Second, tax-free fuel means the industry does not have to bother to change. If aviation fuel were to be a taxed, the cost would fall on airlines and provide a huge commercial incentive to reduce the amount of fuel they burn. Such a tax would put pressure on the whole aviation industry to come up with a more fuel-efficient model for air services. Tax-free fuel traps the aviation industry in the twentieth-century model.

Third, whilst aviation fuel remains tax free there is very little prospect of persuading passengers to embrace clean efficient flying. The reason is that a sustainable future for aviation requires offering passengers a different service. The key insight is that flying fast is inherently energy-intensive and requires a lot of fuel. Although jet aircraft are slowly getting more efficient, the only way to make a real difference is a step change in technology. This leap forward is based on slower air vehicles. The technology for relatively slow efficient air vehicles is well understood but there is no market. Few passengers, apart from committed environmentalists, would choose a relatively slow flight for the same price as a fast flight – even though a slow air vehicle could be more spacious and a better travelling experience. It is also convenient in the context of the 2020 COVID-19 crisis that social distancing should be easier in more spacious cabins. This would reduce the risks of person-to-person transmission of disease, thereby encouraging people back to taking flights as the pandemic recedes.

As I explain below, the obvious sustainable model for aviation is to put significant tax on aviation fuel, making fast flying expensive. Such a tax would not drive up the cost of operating a new breed of relatively slow efficient air vehicles. So, alongside making fast flying expensive, significant taxation of aviation fuel would support developing and deploying a new operating model which is slower but also spacious and comfortable and at a similar price to current fast-jet flying options. My research has shown that generally people would accept this; whereas they would oppose simply making flying expensive, so those least able to pay were priced out of the market.

Whilst aviation fuel remains tax-free, the technology of low-carbon flight is not commercially attractive, so is not being developed, and passengers have no idea that there could be an affordable but slower way to fly. Without an affordable option, governments are not going to make fast flying expensive. Tax-free aviation fuel is a total block on progress. The action required to unlock the industry is deceptively simple; it is to tax aviation fuel. Tough love perhaps, but that is what the industry needs.

# A sustainable model of aviation

The twenty-first-century model of green sustainable aviation is quite different to the old twentieth-century model. Currently, the fast jet is the workhorse of the sky. The aircraft for First Class, Business Class, Economy, and air freight will vary by internal configuration and are differentiated by levels of service on board, but the basic aircraft is a version of fast jet. This will change as alternative low-emission air vehicles become part of the new model of aviation services.

The development of aviation post-World War Two is particularly interesting and helps to explain the trap the industry is in. Speed had been crucial to winning the war in the air and relatively small speed gains could provide the winning edge. The jet engine provided such an edge. After the war, it seemed obvious that this modern technology would have civilian application. The first jet airliner was the British-built Comet. It was a fuel guzzler and unreliable, suffering three crashes in its first 12 months of operation. This was not going to stop the jet age from taking off; the jet airliner was cool and the jet age was the future.

During the 1950s, operating in parallel with the early jets, were propeller-driven airliners. An example was the Russian-built Tupolev Tu-116. It was reliable and safe, but at the dawn of the jet age it was certainly not cool. It was also fast (cruise speed 500 mph; 800 km/h), but not quite as fast jet airliners. Current generation jetliners cruse at around 575 mph, so about 15% faster than the Tupolev Tu-116. The main difference between the Tupolev Tu-116 and the Comet (apart from safety) was fuel efficiency. The Comet's jet engines were a very thirsty early design before engi-neers worked out how to burn fuel efficiently to produce thrust. The Comet has been replaced by better jet aircraft but it has remained useful to the industry as the baseline for reporting jet aircraft efficiency in a favourable light.

The industry has been able to report steady progress over the last 70 years but the really interesting observation is that it is only the newest most modern jets which can match the fuel-efficiency of the Tuplov 116. How can we congratulate ourselves on progress in aviation when, on fuel-efficiency measures, we are still back where we were in the 1950s? What have we been doing for the last 70 years? The answer is that fuel has been so cheap that fuel-efficiency has been low-down on the list of design parameters. Jet engine manufacturers have made good progress but are perhaps approaching the limit of maximum fuel efficiency. The design parameter we should be looking at is the whole system of air services and how they are delivered. Airlines have large fuel bills because they rely on the fast-jet workhorse. So, it is the fast jet which needs to be reconsidered. Perhaps it is not cool any more. Perhaps a slower low-emission air vehicle could be the new cool way to fly.

The current refusal to countenance radically different, highly efficient aircraft is deep-rooted. I had a conversation with a professor of engineering at one of the UK's leading universities. I asked why so little effort was going into developing radically different aircraft (I should use the term 'air vehicle' as they may not look like, and need not look like, current aircraft). The reply I received was that they had examined such air vehicles and concluded that they are not feasible. I pressed for a fuller reply because I knew, that from an engineering perspective, such designs are feasible. I then teased out that the reason they are not considered to be feasible is because they would not be commercially viable. This meant that there was little funding available for such engineering research. Even expert engineers were getting caught in the trap of economic barriers. Radical changes in air services are thought not to be possible because the economics don't work, rather than anything to do with engineering limitations.

The key to a sustainable model for aviation is that not all people nor all freight need to travel fast. Travelling fast will always be energy intensive because air resistance increases exponentially with speed. For example, as a rough estimate, flying at 600 mph will require approximately 40% more energy than flying at 500 mph. Halve the speed from 600 mph to 300 mph and the energy required could be a quarter of that needed to maintain 600 mph. These are significant savings. The maths is not quite so simple in practice, because the design of air vehicles for different speeds can vary considerably, but the principle to apply is that slower can be much more efficient.

I don't want to get into a detailed game of second-guessing what aeroengineers will design for our future air vehicles; nor will I engage in predicting the exact structure of a sustainable model of sustainable air services. However, the approximate shape can be sketched out. Flying fast will be expensive; flying slower will be the affordable option. I don't think that it will be an us-and-them model of air services, in which the rich fly fast and the poor fly slow. The future for time-poor people could be supersonic but it will be a small, very expensive niche. The new cool way to fly for both rich and poor could be on hybrid air vehicles which fly like a plane but also get part of their lift from large spaces filled with the inert gas helium. Although slow relative to jets, these would be large and spacious. First Class could be luxurious; Business Class an excellent office; and Economy passengers would at least have space rather than be crammed in like sardines. It might only be busy business executives and world leaders on a tight schedule who would be uncool and travel in fast jets.

When slower efficient air vehicles enter service the political context changes. Currently such aircraft are not commercially viable because few people would choose to fly slower for much the same price as a fast jet. A better travelling experience is a reason to fly slow, but price would be a more powerful incentive. At the moment, to put a significant tax on aviation fuel would be difficult for politicians because it would hike the price of flying for everyone. The general public would object to their ability to fly being curtailed whilst the better off could carry on as before. When slow fuel-efficient air vehicles have been developed and deployed on most routes, taxing aviation fuel would make the fast jet expensive. Instead of people being priced out of flying, people on a budget would need to choose a slow flight. The better experience, and the cool factor which airline marketing professionals will be able to generate, could overcome resistance. If the rich are seen to pay a lot for speed but the rest of us can still fly, it would become politically acceptable to ramp up tax on aviation fuel.

## The transition

I have had discussions with a wide range of people, from politicians to executives within the aviation industry, about this strange anomaly that aviation fuel is not taxed. It is widely accepted that it is out of date and should be changed but there is also certainty, particularly amongst aviation industry executives and aviation policymakers, that Article 24 cannot be changed. Opposition by the United States is cited as the reason. In the UK, I have spoken with officials and politicians who agree that action is long overdue to remove the tax-free exemption for aviation fuel but see little point in expending effort and political capital trying to persuade the United States. In the absence of demands for change from the UK electorate it does not rise up the policy agenda. So, we remain stuck in the past. Even environmentalists join this charade by seeking to persuade people not to fly rather than complaining about the lack of tax on aviation fuel. In the UK, half the cost of filling up the car fuel tank is tax. The airlines pay zero fuel tax. We should be outraged. Airlines complain that their fuel bills are already huge and they could not afford to pay a significant tax. They say such a tax would decimate the industry. Their fuel bills are high for a reason. It is because they burn huge quantities of fuel. The solution is to burn less fuel, not resist sensible taxation.

The aviation industry is not going to volunteer for the transition to sustainable aviation. Aircraft are manufactured, sold or leased, and operated for 30 years or more. Once the prospect of a transformation looks likely, current aircraft fleets will plummet in value. Investors will examine the balance sheets of airlines and their shares will be marked down. Aircraft manufacturers will be particularly hard hit. The most efficient current jets such as the Boeing Dreamliner should have a continued life, but older less efficient jets will be off to the scrapheap, and orders for new jets dry up. There will also be huge opportunities for forward looking businesses who anticipate the transformation and invest in the development of fuel-efficient relatively slow air vehicles. The capability to design and build such 'aircraft' resides within the big current players such as Boeing and Airbus, but the negative hit to sales of existing models could tip them into bankruptcy. I expect the businesses which arise out of such insolvencies to take up the running. It is not surprising that current management and shareholders are not keen on an early transition to sustainable aviation. The timing is not under their control. When it happens will be dictated by public demand (that means us) forcing politicians to act.

Persuading people to support a transition to sustainable aviation is a difficult sell. In this highly polarized debate people tend to be either for or against flying. It is widely thought of as a binary choice. We can fly and ignore the environmental consequences; or choose not to fly to save the planet. This is a false dichotomy as there is a third way, choosing to fly relatively slowly with a relatively low environmental impact. I am very positive about the future for aviation if we can all agree that the current fuel-guzzling model of aviation should be replaced. The task for world leaders is relatively simple. They need to agree to convene another convention on civil aviation tasked with putting in place new rules for twenty-first-century aviation. There needs to be clear direction to remove the tax-free status of aviation fuel. Ideally a minimum fuel tax rate would be set which escalates over time according to a firm timetable. This would provide a solid and transparent policy basis for aircraft manufacturers and airlines. The complaints from both the industry and many passengers at the disruption which would follow are the cost of holding onto an outdated aviation model for so long. Exactly what emerges we cannot know for sure, but we can be certain that such a change of policy would set off an unstoppable sequence of events which launches aviation into a new era.

# Conclusion

I selected aviation for the penultimate chapter of this book because to deal with aviation is the barometer of when the world is ready to face up to climate change. The action required is deceptively simple. All that is required of governments is to convene an international convention on civil aviation tasked with agreeing provisions to enable the taxation of aviation fuel. This should not be a peppercorn tax which plays lip service to the idea, but a tax at least as big as that applied to fuel for ground transportation. The higher the tax, the more rapid the transition, and the sooner aviation lands in a better place.

No doubt there will be considerable discussion about who collects and keeps aviation fuel tax. That will be for the convention to decide. My view, to ensure governments' support, is that the tax is collected and retained by national authorities, so that national treasuries reap full advantage. Countries with twenty-first-century ambition might use the tax receipts to invest in becoming world leaders in the new aviation technology. To take such a lead could be a huge boost to secure an advanced future-proofed economy. Other countries might simply welcome the additional tax income and continue to buy aircraft manufactured elsewhere.

Finding the resolve to take action over aviation would be a strong demonstration that the world is finally facing up to climate change. All it would take to make a start is widespread public demand for action directed at our politicians. In response, politicians should get together at world level to convene a convention on civil aviation to update the rules for flying in the twenty-first century. That should not be too much to ask of them.

*Face down fossil fuel defenders*

# CHAPTER 11

# GRASP THE OPPORTUNITY

Future generations will not look kindly on those with power and influence who continue to defend an obsolete economic model based on fossil fuel. Our leaders know exactly what is going on with the climate; yet nothing substantial is being done about it. We need to challenge spineless politicians, and refuse to allow those with a vested interest in the status quo to have their way. In writing this book, I do not claim to have crafted a complex, clever solution. The solution is in fact surprisingly simple. It is to close down the fossil fuel economy. All I have done is to expose the simple reality of our predicament. Get real, and we can make progress.

I have written this book due to my concern at the damage we are doing to planet Earth and the knock-on consequences for society. I feared that my message would be sidelined, as just another lone voice, with my words falling on deaf ears. I was wrong to be fearful. Recent developments are changing attitudes. The COVID-19 crisis has reminded people of what is essential in life, what really matters, and what we can do without. Closing down the economy has not been all doom and gloom. People have seen that cleaner and quieter cities are better places to live. We now appreciate skies without the noise and pollution of fast jets. In due course, we will still want to fly, but how much better to be able to travel a little slower on a new type of air vehicle which is cleaner, quieter, and has space for social distancing. To address climate change, a downturn in the current fossil fuel economy was inevitable eventually. COVID-19 has got there first. There is now a wonderful opportunity to rebuild a broken economy into a vibrant new economy redirected away from dependence on fossil fuel.

## Civilization at risk

There are times in the flow of human history when progress takes a nosedive. Seemingly invincible civilizations, such as the Roman Empire, expand and generate enormous wealth, and then collapse. In the chaos of a disorderly downward spiral, sound administration breaks down and accurate records are no longer maintained. Historians are left to guess exactly why the collapse occurred. Why did people at the time not understand what was going on, and take action to stop it? We live in the largest, most powerful civilization there has ever been, extending to all corners of the planet. We are all connected and interdependent. The potential Achilles heel is that we all share one characteristic – an addiction to fossil fuel. It is this addiction, affecting us all, which could bring down the whole world order.

The modern world is so integrated and intertwined that escaping collapse would not be easy. When the world descends into chaos – as it will if we don't take action – nowhere on the planet will be safe. For those people who still believe that climate change will not affect them and is not their problem, consider this. If climate change is the trigger of the collapse of our current global order – as it might well be – there will be knock-on consequences such as nuclear arsenals which are no longer under the control of stable governments. There will be no safe place to hide. Living in a compound protected from climate refugees, located on high ground above rising sea levels, in storm-proof buildings in a region where agriculture can continue will not be enough. Nowhere on the planet will be safe from the fallout. These are dark words, and I hope we are not on the verge of such a breakdown. The only way to make sure I am wrong is to get real and remove the risk.

Fossil fuel addiction is not a threat to planet Earth. The planet will survive this human onslaught, but it will shift to a different climate. The threat is to us. Human expectations to carry on living in the same place, and grow the same crops, will no longer be secure. Finding a safe place to live as communities are forced to relocate is far from certain, even for rich and powerful people. Society is not safe because of police forces or strong armies but because of the unwritten pact that we rely on each other. The give-and-take of living in a complex world has played out in a climate which for centuries has changed little.

Communities, places, and countries evolve to suit the prevailing climate. Agricultural capacity could be relied upon. Although there would be good and bad harvests overall the ability to farm was a stable factor in people's lives. Climate change takes away this stability. Areas which currently have good agricultural land could become deserts. This is not a problem which will affect only people in poor regions of far-flung places; it could affect much of southern Europe. Some of Europe's prime agricultural land may have to rely on irrigation from groundwater. As rainfall becomes increasingly rare, the boreholes could eventually run dry and agriculture will no longer be feasible.

Rising sea levels will mean abandoning many of the of world's greatest cities and moving to higher land. If you already live on higher land, believing you will be safe; think again. You would be safe from the encroaching sea, but you will face pressure to accept people displaced from low-lying coastal areas. It is not fanciful to suppose that amongst this climate-induced turmoil civilized society is at risk of collapse.[49]

---

49  McManners, P.J. (2009), *Victim of Success: Civilization at Risk*, UK: Susta Press.

Climate change will mean that everything on which we rely becomes uncertain. It is like moving civilization into a tumble dryer and seeing what happens. There would be no way to predict the outcome, as everything on which we depend is jumbled up and reordered. All the old certainties would be gone.

Whilst the worst consequences of climate change are not seen directly, we can pretend ignorance, but this won't last long.

Allowing history to move forward, choosing to remain oblivious of the likely consequences, will end up hitting us hard. The consequences will ripple through every nation, rich and poor, impacting everyone inhabiting our shared planet. Everyone is in the frame to be a victim; no one can sleep easy in denial of the truth. Everyone reading this book is responsible and we will all be its victim.

## The blame game

Try as we might to pass the blame to chief executives of oil companies, drivers of gas-guzzling SUVs, or presidents of countries in denial, all of us live in a society which is based on fossil fuel. We are all responsible for climate change. Most executives working for fossil fuel companies, policymakers and politicians are good and honourable people. They should not be singled out for blame or made accountable. All of us are guilty of failing to speak up and take action within our sphere of influence – whatever that might be. Everyone whose circumstances mean they are able to read this book are responsible, because we live within industrialized societies reliant on fossil fuel. None of us can claim not to know; none of us can claim it is not our fault; none of us can claim innocence. We are all responsible – and we are all liable to be victims in one way or another. Instead of sitting idly by observing, we need to find the resolve to do something about it.

This is not a fight against an evil conspiracy, but a fight against ourselves. The fight is within society to make the case that addiction to fossil fuel is dangerous; to accept that real robust action is required; and insist that our leaders get on with it. Of course, it would be easier to believe that it isn't happening; or it will not affect me; or when it does affect me, I can afford to move to a safe location. Focusing on self-protection; to move away from low-lying areas; build storm-proof buildings; and put in place robust arrangements to secure food supplies; are all well and good. Such measures will become increasingly necessary, but open your eyes and take in what you see on the news; listen to what the scientists are telling us; and speak out to say *STOP*, enough of this procrastination. Fossil fuel addiction is a clear and present danger to the world, and the cohesion of society is at risk.

## Taking action

As we allow $CO_2$ levels to continue to rise, the things which really matter, which makes it possible to live a good life, are put at risk. There are measures we can take. We can ensure our house is well protected from extreme weather; we can secure access to food; we can move out of communities located by the sea; but all of us will lose the security of safe and stable societies, as the world is awash with desperate climate refugees. They may not be bad people, but desperate people do what they can to survive. Climate refugees will not only be desperate but have every justification to be angry, especially against those who have relied on fossil fuel for so long. We might hope that such desperate and angry people will meekly accept their fate, but there is no reason why they should. If we don't stand up against fossil fuel addiction, the victims of climate change will do it for us in a way of their choosing. We should not expect to be treated leniently. If we face up to climate change without further delay, there is still a chance that instead of being a downturn in history when it all goes horribly wrong, it could be a step change up when the world strides forward with confidence.

There is a bright and vibrant future beyond fossil fuel addiction but there is no escaping a period of cold turkey. The transformation will be less traumatic the sooner it starts. Whilst there are still ample reserves of gas, these can be used to substitute for dirty fossil fuels such as coal. We can't afford to wait until reserves of such cleaner fuels run low. It is too dangerous to remain stuck in the fantasy world of the addict, committed to the next fix and without the will or the ability to rescue ourselves from our self-inflicted fate. I am fully of the belief that the world is now ready to take the leap forward to face up to climate change. However, my belief is not consistent with the facts of the current reality. The delegates of the UNFCCC have not made progress; our politicians are not engaged in searching for a solution; and the general public are not demanding that they do. For my belief to become reality requires others to join with me in believing that it can be done.

## A momentous decision

Facing up to climate change is the most momentous decision of our age. Why we don't is a puzzle. If our house was under water, or our family starving from lack of food, or our community overwhelmed by climate refugees, and violence was entering previously safe neighbourhoods, we would demand that politicians deal with the cause. These are all possible consequences of the future we are entering. Our inability to face up to climate change seems to come from innate human nature. We tend not to take difficult decisions requiring dramatic action until circumstances force us to do so. If we wait until climate change is ripping apart society, it will then be too late. The main impacts of climate change may be decades away but the cause is here and now. The cause is addiction to fossil fuel. We know that to be so and still we delay.

I could be accused of pressing the analogy with drug addiction too far, but it accurately and precisely describes our current behaviour. The euphoria of living well in society depending on fossil fuel is blanking out the long-term consequences. The idea that the drug should be withdrawn before our health deteriorates is not an idea for today. It seems fine; I feel fine; the scaremongers exaggerate. I need the drug and no one is going to take it off me. We vehemently defend our lifestyle despite overwhelming evidence that we are on the road to ruin. Thinking about fossil fuel dependency in this way, comparing our behaviour with a heroin addict, is like looking in the mirror. We know that long-term dependence on fossil fuel will be bad for our health but we fail to look beyond the timescale of the next fix. No one can say exactly when heroin addiction kills you; but you can be sure it will. You would also be sure, if you could take respite from the influence of the drug, that the longer you wait the more severe will be the withdrawal symptoms; and the more difficult it will be to find the resolve to escape the dependency. Pulling back from a self-centred short-term view allows us to see plainly the situation, bringing us face-to-face with reality.

## Escaping fossil fuel dependency

We have become totally reliant on cheap energy from fossil fuel. Over two centuries the fossil fuel economy has developed and grown. To withdraw fossil fuel now, immediately, would collapse the current economy. It is relatively easy to make the case for the withdrawal of fossil fuel, but this requires building a different economy, and that will take time. There are many people, clever people, well-qualified people, and some economists, who argued that the economic hit would be too great to tolerate. Let us be clear; there will be a short-term economic hit. We are now so deep into fossil fuel dependency that withdrawal will be painful. Even though there is widespread resistance to withdrawing from fossil fuel, the logical argument to do so is cast-iron. The challenge is to accept the necessity of withdrawing from fossil fuel, and accept dismantling the fossil fuel economy. This is the only realistic way to make the space to build the new economy. The challenge is plain; the difficulty is obvious; and the need to get going is ever more urgent. We can either sit tight and allow wilful ignorance and denial to take hold or decide to step up to action.

We must stand up to self-interested parties who resist. We must insist that our politicians step forward and earn their salaries. I would like to say that further delay is unthinkable, but further delay seems to be on the cards. It would be logically, morally, and ethically wrong to carry on regardless. Of course we should act, and use our intelligence and ingenuity to do so. Homo Sapiens replaced the bigger and stronger Neanderthals because clever cooperative behaviours, working as a tribe, won out over the stronger species. Climate change deniers and fossil fuel apologists are currently strong; but through clever cooperative behaviour we can win out to ensure we move into the future we want rather than stagger forward under the yoke of brute denial. We should think clearly, accept the outcome of analysis, and then cooperate to deliver the solution. This is how human societies thrive, so let us get on with it.

## Changing attitudes

The first change required to open up the possibility of action is a shift in attitude. In the modern media age this can be rapid if an idea starts to gather momentum, attracting supporters and expanding to become a massive shift in sentiment shared by many people. This doesn't happen with reports by academics and civil servants written carefully so as to avoid offence and explaining only as much as sponsors are happy to read. Such reports might be accurate, containing only those insights for which there is robust evidence and including nothing controversial or offensive to vested interests. Such carefully calibrated reports will not generate widespread demand for change.

To get going with facing up to climate change we need a sense of urgency, and a change of direction. We have to accept that the UNFCCC process cannot, and will not, deliver the solution. Whilst there is belief that it can, this gives policymakers and political leaders cover to justify doing very little. There is even a dangerous idea, which occasionally surfaces, that we should not do too much too soon, to retain the capacity to make cuts to comply with any future agreement. Continued belief in the UNFCCC also ensures that dissent from the wider population is deflected. Demands for action become demands for greater support for the work of the UNFCCC, rather than demands that it should be scrapped and replaced. Good and honest people are being persuaded that the UNFCCC is the forum to target, to persuade their national delegation to be bold. I have a message for such good and honest people – the UNFCCC is not the solution, it has become a distraction. It was set up with good intentions but it has achieved all it can and it is time to move on. Its positive contribution has been to provide solid evidence of how much more carbon dioxide can be emitted before the consequences become serious. The UNFCCC now needs to be downsized to focus on keeping the climate science under review. The idea that the UNFCCC might broker the solution should be abandoned. There is little point in waiting for the moment in the future when it is becomes abundantly clear that the UNFCCC has failed. We would then regret that we did not accept this reality sooner. Waiting until there are no other choices other than to do something may be hard-wired into the international bureaucratic system, but it can be bypassed.

Widespread public concern could bounce world leaders into responding. This could include protests, demonstrations, and civil disobedience. If world leaders continue to ignore demands, public concern might go beyond protest and complaints to include taking direct action. There is little each of us can do individually, but everyone calling for change can become an irresistible force. One person calling for change might persuade two or three other people to call for change, who then persuade more people to call for change, growing into a huge collective demand for action. For protest to expand in this way requires an elusive and hard to define kernel of truth which touches people's core beliefs and identity. The climate science should be enough but that may be too dry and academic for popular appeal. I have to admit that I don't know what the key truth will be, and it may be different for different people – severe storms, flooded cities, the threat from climate refugees, the extinction of species or any one of the many negative consequences of climate change. The avoidance of negatives might do it, but the kernel of truth which tips us into action may be insights into how much better it could be. The COVID-19 crisis exposed the risky nature of global interconnectedness as the world economy closed down. It also gave people a glimpse of how much cleaner and more pleasant cities could be, and reminded us what is essential, what we value, and what we can live without. This could help persuade people that the changes required are a huge leap forward to a better society supported by a different and sustainable economy.

# Knowing what to demand

What I do know, and can communicate here with clarity, is what needs to happen next. It is obvious, but easily overlooked, that we need to know what to demand if protest is to serve a purpose. It may not be possible to know exactly what will trigger widespread concern, or how it will finally grab people's attention to inflate into a massive demand for change, but unless there is a plan of action it could deflate just as quickly. We have seen this happen a number of times already. The climate talks in Copenhagen 2009, and Paris 2015, or the outspoken young person Greta Thunberg in 2019. These all generated huge publicity, demonstrations and public concern, and demands for action. But without knowing what action should be demanded these bubbles of protest do not endure for long. Insights into a different possible future from the 2020 COVID-19 crisis could evaporate just as quickly. The world returns to business as usual until the next press report, of a major city flooding, glaciers melting or risk of polar bear extinction, hits the headlines.

For protest to be effective, there needs to be a clear agenda for the action being demanded. As the protest gathers steam, political leaders and policymakers will then know what is required of them. Protesters will also know whether their demands are being met. Entering political protest without a clear agenda risks being fobbed off with words of reassurance that something is being done, without the politicians being called to account.

The agenda I put forward is clear and cuts through to what will have to happen eventually. Before getting angry and resolving to join the protests, I urge you to become familiar with this agenda.

We should demand a real pragmatic solution to climate change. We should accept that seeking to control carbon emissions has not worked, and is unlikely to work. Whilst fossil fuel is readily available, we remained trapped by our addiction. We are failing to find the discipline to close down our fossil fuel habit. Instead of trying to cure the addiction through persuasion which falls on deaf ears, the cure is to be found in closing down fossil fuel extraction. The key insight is that to burn all known reserves of oil and gas will smash the budget for staying within 1.5°, and take us most of the way to the upper limit to avoid dangerous climate change of 2°. This leads logically to closing down the dirty fossil fuels, particularly coal but also oil extracted from tar sands and shale rock. Society is then forced to rely on conventional oil and gas for the decades ahead. We will have no choice but to change how the economy operates. This will be so much more effective than hoping to close down fossil fuel through voluntary choice. Commitments to use less fossil fuel can easily be reneged on. It would be much harder to open new facilities for extracting coal when there is an international treaty preventing it. Even if the addict wanted to return to their old habits, investors and banks would refuse to provide the funds. This approach can deliver a closed and effective solution.

To summarize what we should be requiring of our politicians, there are three demands:

1. A global treaty on closing down fossil fuel extraction focused on coal and other less efficient fossil fuels.

2. Trading arrangements for the cleaner fossil fuels, gas and oil, to manage the transition.

3. A new convention on civil aviation tasked with establishing a mechanism to tax aviation fuel to demonstrate our resolve that we will face up to climate change.

These are doable, precise, and focused where maximum impact can be delivered for the least effort. Let us not obscure our demands with all sorts of other related matters, no matter how well justified. All of us, sharing the same determined attitude, and the same agenda, can make it so. Let's keep it clear, keep it simple, and demand that it happens.

# REFERENCES

Bowen et al. (2015) 'Two massive, rapid releases of carbon during the onset of the Palaeocene–Eocene thermal maximum' *Nature Geoscience* 8, 44-47. [Available online: https://www.nature.com/articles/ngeo2316; accessed 26 July 2020].

British Petroleum, (2019), *BP Statistical Review of World Energy 2019*, 68th edition [bp.com/energyoutlook].

Centre for Alternative Technology (2010), *Zero Carbon Britain 2030*, M. Kemp and J. Wexler (eds), UK: CAT Publications.

Centre for Alternative Technology (2017), *Zero Carbon Britain: Making it happen*, editors C. Toms, A. Hooker-Stroud and A. Shepherd (eds), UK: CAT Publications.

Cummins, N. (2019), 'China Plans to Build an Astonishing 200 New Airports in the Next 15 Years', online article, 16 January, [https://simpleflying.com/china-to-build-250-new-airports/; accessed 20 May 2020].

Elkington, J. (1997), *Cannibals with Forks: The Triple Bottom Line of 21st Century Business*, UK: Capstone.

Environment Agency (2012), *Thames Estuary 2100, TE2100 Plan*, November.

Farman, J.C., B.G. Gardiner and J.D. Shanklin (1985), 'Large Losses of Total Ozone in Antarctica Reveal Seasonal ClOx/NOx Interaction', *Nature*, 315, 207–210.

Goldfrank, W.L, A. Szasz and D. Goodman (eds) (1999), *Ecology and the World-system (Contributions in Economics & Economic History)*, USA: Greenwood Press.

IATA (2018), 'IATA Forecast Predicts 8.2 billion Air Travelers in 2037', Press Release 62, 24 October, [https://www.iata.org/pressroom/pr/Pages/2018-10-24-02.aspx; accessed 5 Nov 2019].

IATA (2019), 'Destination Green; The Next Chapter', Environment Report.

ICAO (2006), Convention on International Civil Aviation – Doc 7300/9 Ninth Edition, [https://www.icao.int/publications/Documents/7300_cons.pdf; accessed 13 Nov 2019].

IPCC (2014), *Climate Change 2014: Synthesis Report, Contribution of Working Groups I, II and III to the Fifth Assessment Report of the Intergovernmental Panel on Climate Change* [Core Writing Team, R.K. Pachauri and L.A. Meyer (eds.)], Switzerland: IPCC.

IPCC (2019), Global Warming of 1.5°c. An IPCC Special Report on the impacts of global warming of 1.5°c above pre-industrial levels and related global greenhouse gas emission pathways, in the context of strengthening the global response to the threat of climate change, sustainable development, and efforts to eradicate poverty [V. Masson-Delmotte, P. Zhai, H.-O. Pörtner, D. Roberts, J. Skea, P.R. Shukla, A. Pirani, W. Moufouma-Okia, C. Péan, R. Pidcock, S. Connors, J.B.R. Matthews, Y. Chen, X. Zhou, M.I. Gomis, E. Lonnoy, T. Maycock, M. Tignor, and T. Waterfield (eds.)], [https://www.ipcc.ch/site/assets/uploads/sites/2/2019/06/SR15_Full_Report_High_Res.pdf].

Le Quéré et al. (2018), Global Carbon Budget 2018, *Earth System Science Data*, 10, 2141–2194.

Le Quéré, C., Jackson, R.B., Jones, M.W. et al. (2020), Temporary reduction in daily global $CO_2$ emissions during the covid-19 forced confinement, *Nature Climate Change,* online 19 May, [available from: https://doi.org/10.1038/s41558-020-0797-x: accessed 19 May 2020].

Lynas, M. (2007), *Six Degrees: Our Future On A Hotter Planet*, London: Fourth Estate.

McManners, P.J. (2008), *Adapt and Thrive: The Sustainable Revolution*, UK: Susta Press.

McManners, P.J. (2009), *Victim of Success: Civilization at Risk*, UK: Susta Press.

McManners, P.J. (2012), *Fly and be Dammed: What now for aviation and climate change?* London: ZED Books.

Randles, J. (2017), 'Judge Rules Peabody Energy Bankruptcy Blocks Global-Warming Lawsuits', *Wall Street Journal*, 25 October. [Available online: https://www.wsj.com/articles/judge-rules-peabody-energy-bankruptcy-blocks-global-warming-lawsuits-1508961807; accessed 23 October 2019].

Ritchie, H. and M. Roser (2019), 'Fossil Fuels', Published online at OurWorldInData.org. [Retrieved from: https://ourworldindata.org/fossil-fuels].

Shepherd, A., Ivins, E., Rignot, E. et al. (2019), 'Mass balance of the Greenland Ice Sheet from 1992 to 2018', *Nature*, doi:10.1038/s41586-019-1855-2.

'COP24: Not all hot air', *The Economist*, 22 December 2018, 120–121.

UNFCCC (2015), Conference of the Parties, Twenty-first session, Paris, 30 November–11 December, document dated 12 Dec 2015 [available from: https://UNFCCC.int/resource/docs/2015/cop21/eng/l09r01.pdf; accessed 21 Oct 2019].

UNFCCC (2018), United Nations Climate Change Annual Report 2018.

UNFCCC (2020)a, Introduction to Mitigation [available from: https://unfccc.int/topics/mitigation/the-big-picture/introduction-to-mitigation; accessed 17 March 2020].

UNFCCC (2020)b, The History of the Convention [available from: https://UNFCCC.int/process/the-convention/history-of-the-convention; accessed 17 March 2020].

Wallace-Wells, D. (2019), *The Uninhabitable Earth: A Story of the Future*, London: Penguin.

# INDEX

**217**